Principles of Oceanography

Seventh Edition

M. GRANT GROSS
Chesapeake Resource Consortium
& University of Maryland

Prentice Hall
Englewood Cliffs, New Jersey 07632

Library of Congress Cataloging-in-Publication Data

Gross, M. Grant (Meredith Grant), 1933–
    Principles of oceanography / M. Grant Gross. – 7th ed.
        p.   cm.
    Rev. ed. of: Oceanography, a view of the earth.
    Includes index.
    ISBN 0-02-347981-7
    1. Oceanography.   I. Title   II. Series: Gross, M. Grant
    (Meredith Grant), 1933–   Oceanography, a view of the earth.
    GC16.G7   1995
    551.46--dc20   94-47992
                                                                    CIP

Cover art: © Hans Strand/Tony Stone Images
Editor: Robert A. McConnin
Text Designer: Ed Horcharik
Production Buyer: Patricia A. Tonneman
Production Coordination: Custom Editorial Productions, Inc.
Illustrations: Rolin Graphics Inc.

This book was set in Meridien and was printed and bound by R. R. Donnelley & Sons
Company. The cover was printed by Phoenix Color Corp.

 © 1995 by Prentice-Hall, Inc.
A Simon & Schuster Company
Englewood Cliffs, New Jersey 07632

Earlier editions, entitled *Oceanography,* © 1990 by Macmillan Publishing Company and
© 1985, 1980, 1976, 1971, 1967 by Merrill Publishing Company.

Printed in the United States of America

10 9 8 7 6 5 4 3 2 1

ISBN 0-02-347981-7

Prentice-Hall International (UK) Limited, *London*
Prentice-Hall of Australia Pty. Limited, *Sydney*
Prentice-Hall Canada Inc., *Toronto*
Prentice-Hall Hispanoamericana, S. A., *Mexico*
Prentice-Hall of India Private Limited, *New Delhi*
Prentice-Hall of Japan, Inc., *Tokyo*
Simon & Schuster Asia Pte. Ltd., *Singapore*
Editora Prentice-Hall do Brasil, Ltda., *Rio de Janeiro*

Preface

Oceanography—the scientific study of the ocean—deals with many aspects of Earth that make it unique among the Solar System planets. Viewed from space, the Earth's blue ocean and white clouds attest to the abundance of water on Earth, compared with the dry surfaces of nearby planets—except, of course, for Venus with its cover of thick acid clouds.

Recent exploration of the Solar System has brought about great advances in our knowledge of Earth and its planetary neighbors, providing a new context for studies of Earth's ocean, atmosphere, and life. Moreover, the end of the Cold War has resulted in several previously classified observing techniques becoming available for civilian studies. These advances, plus the huge amounts of data about the ocean previously held in classified form during the Cold War, have greatly expanded our understanding of Earth and the ocean.

In studying Earth's ocean, we must also remember its close linkage with the oxygen-rich atmosphere, which is powered by energy from the Sun, through evaporation of ocean water, and by warming, primarily at the ocean surface. A major question that arises many times in this edition is how changes in Earth's atmosphere affect change its climate. The ocean plays a large but poorly understood role in controlling such processes.

The ocean obscures one of Earth's most distinctive processes, the continued cooling of its interior and the surface-plate movements and sea-floor volcanic eruptions caused by these cooling processes. As we shall see, these cooling processes are the sources of other unique features.

Then there is life itself. Life began in the ocean, and most living forms have close relatives still living in the ocean. Furthermore, life as we know it, even on land, continues to depend on the ocean.

Because of the increasing awareness of the limits on Earth's resources and the need to protect them for future generations, this book includes discussions of sustainable development and various pollution problems. The emphasis is on the understanding of how ocean processes are involved and how they might be used to remedy problems in the future.

Finally, the book discusses the recently gained understanding of the importance of catastrophes affecting Earth, such as the demise of the dinosaurs about 65 million years ago. The series of comet fragments that struck Jupiter in July 1994 provided dramatic evidence of Earth's vulnerability to such events. The record of such events is recorded in the sediments on the ocean bottom. This book emphasizes how such events and other smaller ones are studied using the records in the ocean. Furthermore, the role of the ocean in possible global warming is thoroughly discussed.

Thus, in learning about Earth's ocean, we are learning about Earth itself and how it influences our very existence.

THE BOOK

This seventh edition, retitled *Principles of Oceanography,* retains basically the same format as that of previous editions of *Oceanography.* Its nine chapters are intended for use either as a stand-alone text for a one-quarter introductory course at either the high-school or college level, or as a supplemental text in other courses, such as Earth sciences, marine geology, marine biology, or the physical sciences. It assumes only high-school-level scientific knowledge and and no mathematics. The emphasis remains on the understanding of scientific principles rather than mastery of a large and unfamiliar vocabulary.

The entire text has been rewritten to emphasize the fundamental processes controlling the ocean, the ocean basins, and oceanic life. Each chapter has been expanded and reorganized, with new text and new figures added. Additions to the text include recent advances in ocean sciences and in the many other branches of science that contribute to our understanding of Earth and its ocean. The chapters have been planned so that they stand alone and can be assigned in any order.

The illustrations have been revised, and new ones have been added. More end-of-chapter study questions and new supplementary readings have been included, and the Key Terms and Concepts sections have been expanded.

WHAT'S NEW

Satellite observations of Earth and our neighbors in space have revolutionized our view of the ocean and greatly expanded our ability to make useful predictions. Therefore, this aspect of ocean sciences has been expanded.

Increased concerns about climatic changes on a global scale and the appearance of holes in the ozone layer above Antarctica and the Arctic Ocean are discussed. The ocean's role in regulating climate and the records of Earth history contained in marine deposits are treated in new sections.

The end of the Cold War gave oceanographers access to powerful observing techniques previously used to monitor enemy submarines. Now these techniques are being used to detect volcanic eruptions on the sea floor as they occur, permitting scientists to study the processes involved in these events without delay. These listening networks are also being used to track the movements of individual whales, and as a result we are learning more about the distribution, abundance, and behavior of whales than we ever knew before. Results of these new findings are also featured.

Recent advances in navigation, communication, and computer sciences have helped revolutionize ocean science. New sections discuss these developments.

Several new features have been added. The first is a table of conversion factors (Appendix 1) to make it easier for students to deal with the metric system. Secondly, Appendix 2 outlines important events in Earth and ocean history. Finally, a glossary of important terms and concepts has been added.

UNITS

The seventh edition uses metric units and includes (in parentheses) the common-unit equivalents. Most graphs include both metric and common units. As noted above, Appendix 1 provides conversion factors for the units that are used most often.

ACKNOWLEDGMENTS

The revision of this edition was greatly helped by the many comments and suggestions from my colleagues who have worked and taught with me over the years. I am especially grateful to Bob Foster, formerly of San Jose State University; Don Lear, Anne Arundel Community College; Jon Sharp, University of Delaware; and Ted Packard, University de Rimouski, Quebec, Canada, for their comments and suggestions in reviewing a draft of this edition.

Finally, I must acknowledge the support of my family—especially my wife Liz, who has patiently and lovingly endured all my struggles to put my thoughts into words and figures.

Contents

1
Earth and Human Activities

Most people live within a few hundred kilometers of the seashore, and all of our lives are influenced by the ocean, which provides recreation and food, receives our waste discharges, and is our global highway. The ocean provides water and heat to the atmosphere, which powers the winds, causes our weather, and controls our climate. By contrasting the large daily temperature changes in inland deserts with the limited temperature ranges in coastal areas, we can see how thoroughly the ocean affects our daily lives.

In short, ours is a water-conditioned existence. By studying the ocean, we learn about life's beginnings in the ocean and how much life continues to depend on it. We also learn about the uniqueness of Earth and the fragility of our life-sustaining systems.

1.1 ORIGIN OF EARTH

To understand how Earth and its ocean formed, we begin with the universe (Figure 1.1), which formed in a gigantic explosion (called the **Big Bang**) about 15 billion years ago. (See Appendix 2 for a simplified time table of Earth history.)

Hydrogen, the simplest element, formed first; next came helium, the next-simplest element. These gases were not distributed uniformly in the universe, but formed enormous gas clouds. These clouds condensed by gravitational attraction to form stars, which are objects massive enough for the pressures in their interiors to reach levels sufficient to initiate nuclear fusion, eventually causing them to radiate energy. Clusters of stars then formed galaxies.

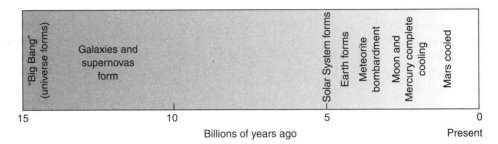

FIGURE 1.1
Schematic representation showing the various stages in the development of the universe, the Solar System, and Earth.

Stars radiate energy formed by fusing of hydrogen, the basic fuel of stars. Eventually each star uses up its hydrogen supply. At this stage the star contracts; the succeeding course of events is determined by the size of the star. Some stars are large enough so that they explode in supernovas, enormous explosions that blow materials from the doomed star into space, forming dust clouds. Formation of other elements occurs primarily during supernovas.

The Sun and the planets orbiting it (the **Solar System**) apparently formed from a dust cloud, also known as a nebula. Again, gravity drew the particles together. Eventually the cloud was rotating rapidly enough to flatten and become disk-shaped. Most matter remained in the center, contracted by gravity, and as nuclear reactions began, our Sun (an average sort of star) was born and began radiating energy.

Elsewhere, the cloud continued contracting and forming smaller clouds within it. Some of these clouds formed planets, taking about 100 million years. Planets forming near the Sun lost much of their original gases, blown away by solar winds. Thus, the inner planets (Mercury, Venus, Earth, and Mars) are known as **terrestrial planets** because they consist primarily of rock. Both Earth and Moon formed about 4.6 billion years ago. The oldest rocks found on Earth's surface were formed about 3.8 billion years ago, after the intense meteorite bombardment ceased. Mineral grains from even older rocks have been identified, but little is known about the first billion years of Earth history.

Soon after Earth formed, its rocks became segregated by their densities. This process probably required many millions of years to complete and released enormous amounts of energy. Eventually, the densest materials (probably iron containing hydrogen) migrated to Earth's center, forming the core. Less dense rocks lie above the core, forming the mantle. The least dense rocks were segregated to form Earth's crust, its outermost shell.

Planets farther from the Sun (Jupiter, Saturn, Uranus, and Neptune) kept more of their original gases, making them much larger but less dense than the terrestrial planets (Figure 1.2). They consist primarily of hydrogen and helium; some may have small, rocky cores. Jupiter is regarded as a failed star. It has the right composition to support fusion, the source of energy for stars, but it is not massive enough for fusion to begin. Pluto, the outermost planet, is little known but is clearly different from its large, gaseous neighbors.

Between Mars and Jupiter is a large belt or band of small rocky-metallic bodies called **asteroids,** which are much smaller than planets and lack the gaseous envelopes that make **comets** so dramatic when they approach the Sun. This belt is the source of the meteorites that still strike the planets, forming impact craters on the surfaces of the terrestrial planets. Studies of meteorites that have struck Earth provide information about Earth's interior composition. (**Meteorites** are thought to be fragments of planetlike bodies that broke up through collisions. Thus, they are assumed to be typical of rocks deep in Earth's interior, not otherwise available for study.)

The Moon and the nearby planets have many conspicuous impact craters. Earth has doubtlessly been struck by many meteorites, but weathering and renewal of its crust have destroyed most traces of them. Some meteorites that have struck Earth have caused catastrophes, including massive extinctions of life, which we discuss later. Jupiter's closeness to Earth partially protects Earth

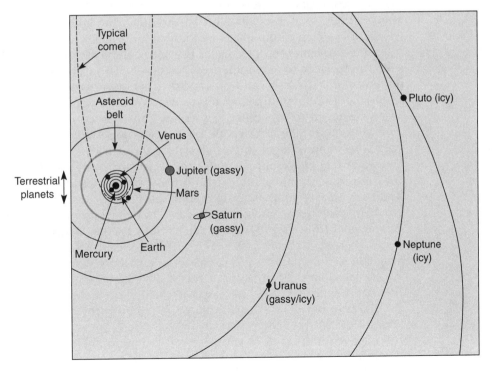

FIGURE 1.2
Diagram of Solar System, showing the positions of the planets. The inner planets are made of rock and are called the terrestrial planets. Planets farther from the Sun have large amounts of gases or icy exteriors. Some may have rocky cores.

from meteorite impacts. Thus, Jupiter suffers more collisions than does Earth. The most recent such impacts occurred in late July 1994. Such impacts on Earth would doubtlessly have caused catastrophic damage and possible extinction of some life forms. The odds against such an event hitting Earth are estimated at one in 50 million.

The Moon's origin is still disputed. One popular theory is that a Mars-sized object struck Earth early in its history. Materials ejected by the impact at first orbited Earth and later aggregated, again by gravity, to form the Moon. In any case, Earth and Moon are essentially twin planets, revolving around a common center. (We learn more about how the Moon effects Earth's ocean when we discuss tides in Chapter 6.)

1.2   ORIGIN OF OCEAN AND ATMOSPHERE

As exploration of the Solar System by satellites has shown, Earth is unique among the Sun's planets, being largely covered by liquid water, having an oxygen-rich atmosphere, and having photosynthesizing organisms, which are able to make their food from sunlight. Furthermore, the limited temperature ranges on Earth's surface permit life as we know it.

The first question to consider is where did the water in the ocean and the gases in the atmosphere come from? When Earth formed, it was hot and dry. Furthermore, strong solar winds must have swept away the planet's original atmosphere and any surface waters as well. We know from studying stony meteorites (assumed to be similar to rocks in Earth's interior) that rocks in Earth's interior contain about half a percent water. Some water was extracted from these rocks and brought to Earth's surface through volcanic eruptions. Some of the water and even some organic matter, such as amino acids (building blocks for cells), came from collisions with comets, which consist primarily of ice, dust, and carbon-based compounds formed in space.

This may not seem like much water, but extraction of only a small fraction of the water from Earth's interior can account for the volume of seawater in the ocean. Indeed if all the water were extracted from Earth's interior, the ocean would be about 20 times deeper than it is, and there would be no land at all. As we see later, some ocean water is recycled back into Earth's interior through its internal cooling processes (discussed in Chapter 2).

Atmospheric gases accumulated together with the ocean (Figure 1.3). Gases released from interior rocks by volcanic eruptions accumulated to form the relatively thin shell of gases we call the atmosphere. After life evolved on Earth, photosynthesis changed the composition of atmospheric gases (atmospheric composition is shown in Figure 3.10 and discussed in Chapter 3). Today, little carbon dioxide remains in Earth's atmosphere, but carbon dioxide dominates the atmospheres on most lifeless planets. Oxygen, released by plant photosynthesis, is a major component of Earth's atmosphere but is absent on the other planets. (We discuss this further in Chapters 4 and 8.)

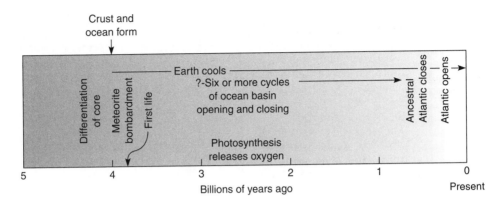

FIGURE 1.3
Schematic representation of the origins of Earth's crust, ocean, and atmosphere.

1.3   DISTRIBUTION OF LAND AND OCEAN

Let's look now at some of Earth's vital statistics.

Earth's radius	6378 km	(3963 mi)
Average ocean depth	3.7 km	(2.3 mi)

Although ocean covers 70.8% of Earth's surface, its depth is insignificant (1/1700) when compared with Earth's radius. (If Earth were a basketball, the ocean would be only a thin film of water and the atmosphere only a coat of paint.) Excluding waters in near-surface rocks, the ocean contains 98% of the planet's free water (Figure 1.4). (Most of the remainder is ice, covering the Antarctic continent.)

Ocean basins and continents are unevenly distributed over Earth's surface (Figure 1.5). Continents and ocean basins are generally opposite each other. For instance, Antarctica, a large continent at the South Pole, is opposite the Arctic Ocean basin at the North Pole. Furthermore, most land lies in the Northern Hemisphere whereas the Southern Hemisphere is mainly ocean.

	Land (%)	Ocean (%)
Northern Hemisphere	39.3	60.7
Southern Hemisphere	19.1	80.9

Ocean waters cover 71% of Earth's surface, and between latitudes 40°S and 60°S there is little land to impede winds or deflect ocean currents. When seen from above the South Pole, the ocean appears as a broad band surrounding the Antarctic continent with three northward projecting basins (Figure 1.6).

There is only one interconnected **world ocean,** but for convenience we divide it into parts: the Atlantic (including the Arctic) Ocean, the Indian Ocean, the

FIGURE 1.4
The ocean contains nearly all of the free water on Earth's surface (excluding water contained in rocks and minerals). About 2% of Earth's water is contained in the ice sheets, primarily in Antarctica and Greenland. (Percentages total more than 100 because of rounding.)

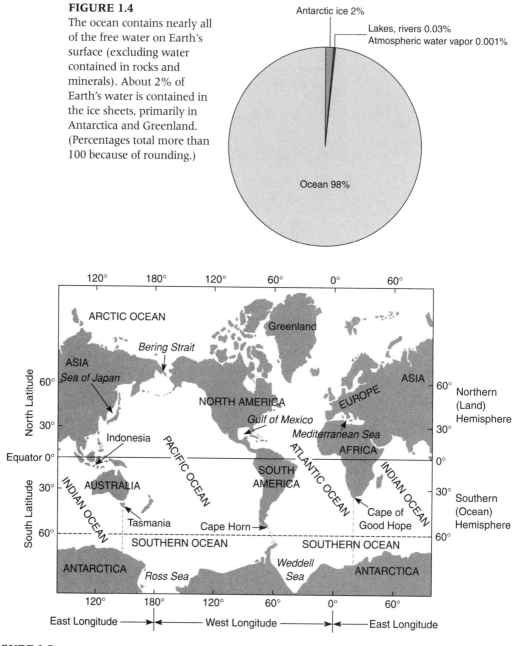

FIGURE 1.5
Continents and ocean basins, showing the boundaries of the major ocean basins. The northern limit of the Southern Ocean is (approximately) 60°S latitude; the southern limit is Antarctica. (Compare view of ocean shown in Figure 1.6.)

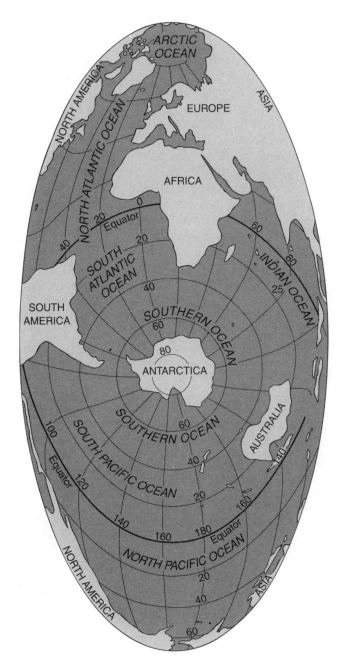

FIGURE 1.6

The world ocean, viewed from the South Pole. Note that the major ocean basins are connected in the Southern Ocean around Antarctica.

[After A. F. Spilhaus, "Maps of the World Ocean," *Geographical Review* 32 (1942), p. 434]

Pacific Ocean, and the Southern Ocean (Figure 1.5). Where land boundaries are absent, we delineate the ocean basins using arbitrary lines. For example, we separate the Pacific and Indian oceans by a line running south, along longitude 150°E, from Australia to Antarctica. The Indonesian islands, north of Australia, form a natural but leaky boundary between the Pacific and Indian oceans. The shallow Bering Strait (between Alaska and Siberia), blocks all but shallow ocean currents, and thus separates the Pacific Ocean and the Arctic Ocean. A line running south from Cape Horn (the southern tip of South America) on longitude 70°W separates the Pacific and Atlantic. A line running south from the Cape of Good Hope (the southern tip of Africa) on longitude 20°E separates the Atlantic and Indian oceans.

The **Pacific Ocean,** the largest basin, is as big as the Indian and Atlantic oceans combined, holding slightly more than half the water in the world ocean (Figure 1.7). The Pacific is the deepest basin; it has few shallow seas but many deep trenches. These trenches are associated with active volcanoes and frequent earthquakes and form the Pacific "Rim of Fire," a belt of active volcanoes and mountain building, which we discuss further in Chapter 2.

The few large rivers discharging into the Pacific Ocean are on the Asian margin—China's Yellow (Huang He) and Yangtze rivers, for example. The Pacific's surface area is ten times larger than that of the land areas that drain into it. Because of its large size, the Pacific is less affected by the surrounding lands than are the Atlantic and Indian oceans. Conditions in the Pacific—especially in the South Pacific—most nearly resemble the conditions one would predict for a water-covered Earth.

The **Atlantic Ocean** is a relatively narrow, twisted body of water bounded by roughly parallel continental margins. Including the Arctic Ocean, the Atlantic has the greatest north-south extent of all the ocean basins, connecting the Arctic

FIGURE 1.7
The ocean covers 70.8% of Earth's surface. The Pacific Ocean contains as much water as the Atlantic and Indian oceans combined.

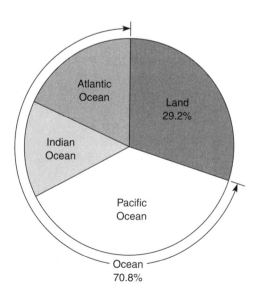

and Southern oceans. Because of its many shallow, marginal seas, wide continental shelves, and the Mid-Atlantic Ridge, the Atlantic is the shallowest of the three major ocean basins (Atlantic, Pacific, and Indian). Many rivers, including the Amazon (the world's largest) and the Zaire (Congo), discharge into it.

The **Indian Ocean** is a Southern Hemisphere ocean, extending only a short distance north of the equator, to about 25°N (Figure 1.5). The northern Indian Ocean is also strongly influenced by large-river discharges (Ganges and Brahmaputra) from the Himalaya Mountains and by seasonally changeable (monsoon) winds. (We discuss the monsoons and their effects on currents in Chapter 5.)

1.4   POLAR OCEANS

The **Southern Ocean** surrounds Antarctica, which forms its southern boundary; its northern boundary is less well defined, although 60°S latitude is often used. The northern limit of the Southern Ocean currents is marked by the Antarctic Convergence, where ocean currents flow together and sink.

The Southern Ocean contains the largest east-west current, the Antarctic Circumpolar Current, which we discuss in Chapter 5. Sea ice forms each winter, covering about 25 million square kilometers at its maximum (usually in September), but about 75% of this ice melts in local summer. Thus, there is little thick multiyear sea ice. The antarctic continent is covered by a thick sheet of ice that has accumulated over millions of years.

Massive ice shelves, formed by glacial ice sheets flowing off the continent, extend out into the ocean. When these glaciers reach the ocean, they break up, forming gigantic flat-topped icebergs, sometimes as large as the state of Rhode Island. Such large icebergs have been tracked by satellites for over a decade as currents moved them. The Southern Ocean connects all three major ocean basins and has been studied since the 1950s under the Antarctic Treaty, which maintains Antarctica as a preserve for scientific research.

The **Arctic Ocean** is a northern extension of the Atlantic Ocean, quite unlike the Southern Ocean. It is nearly surrounded by land, connecting with the Atlantic between Greenland and Norway. Much of the Arctic Ocean is covered by a permanent ice cap, 3 to 3.5 m (10 to 11 ft) thick, which thickens to 4 m (13 ft) at the end of winter, but melts back to about 3 m (10 ft) at the end of summer. Pressure ridges, formed by winds piling ice sheets on top of each other, obstruct movements of icebreakers and submarines moving below the ice. (Such pressure ridges are much less common in the Southern Ocean, because sea-ice movements are relatively unconstrained by land, unlike the Arctic.)

1.5   OCEAN CHARACTERISTICS

Because ocean basins interconnect, processes acting in the most remote basin eventually affect the entire ocean. One illustration of this is the effect of the

Mediterranean Sea on the Atlantic. The Mediterranean's arid climate causes extensive evaporation, making its waters saltier than the adjacent Atlantic. Because of the narrow connection between basins (Strait of Gibraltar), warm, salty waters from the Mediterranean can enter the Atlantic Ocean. They can be detected about 1.5 km (nearly 1 mi) below the surface over much of the North Atlantic. This injection of salty Mediterranean waters helps make the North Atlantic the saltiest of the major ocean basins. (We will see later how this affects deep-ocean currents and Europe's climate in Chapter 5.)

Because of the ocean's age and slow rate of change, most constituents in seawater are well mixed. The composition of salts dissolved in seawater have not changed markedly for hundreds of millions of years. Bottom waters in deep-ocean basins return to the surface in 500 to 1000 years; thus, ocean waters are thoroughly mixed and the chemical composition of the ocean is virtually constant everywhere.

1.6   SURFACE ELEVATIONS AND DEPRESSIONS

Earth's surface exhibits two dominant levels (Figure 1.8):

Land elevation (average):	840 m (2750 ft) above sea level
Deep-ocean bottom:	3730 m (2.32 mi) below sea level

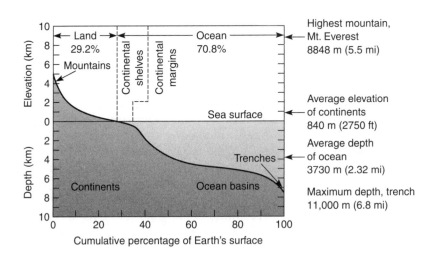

FIGURE 1.8
Hypsographic curve showing the percentage of Earth's surface above any given depth or altitude.
[After H. U. Sverdrup, M. W. Johnson, and R. H. Fleming, *The Oceans: Their Physics, Chemistry, and General Biology* (Englewood Cliffs, NJ: Prentice-Hall, Inc., 1942)]

These two levels are manifestations of differences in density between the two types of rock—one forming continents, the other forming ocean bottom. Continents, including the land, consist primarily of less dense granitic rocks, rich in silica and aluminum. Ocean basins are underlain by denser basaltic rocks, rich in iron and magnesium.

Earth's crust floats on the underlying, partially liquid rocks (discussed in Chapter 2). Continents float higher than ocean basins because granitic rocks are less dense than basaltic rocks. At depths less than 1000 m (3300 ft), the ocean floor rocks are usually continental; at depths greater than 4000 m (13,000 ft), the ocean bottom is normally basaltic.

Mountains higher than 6000 m (20,000 ft) are rare (Figure 1.8). So are parts of the ocean floor deeper than 6000 m (20,000 ft). These extreme heights and depths occupy less than 1% of Earth's surface.

1.7   STUDYING THE OCEAN

Instruments lowered from ships (Figure 1.9) are commonly used to study ocean features and processes. Until the 1950s, oceanographic expeditions were supported by individual countries, as was the nineteenth century, British-supported

FIGURE 1.9
The research vessel *New Horizon* is operated by Scripps Institution of Oceanography, University of California at San Diego. Such vessels provide laboratory and living space for 10 to 20 scientists and their analytical equipment and computers. They can endure intense storms at sea and travel thousands of kilometers without refueling.
(Courtesy Scripps Institution of Oceanography, University of California, San Diego)

Challenger Expedition (1872–1876). This converted British Navy vessel made the first global investigation of the ocean, marking the beginnings of modern ocean science. It sampled all the major ocean basins; discovered the continuous nature of the Mid-Atlantic Ridge; and systematically studied the deep-ocean currents. The 50-volume *Challenger* Expedition Report remained the primary source of information about the global ocean until the 1950s.

The large international expeditions that followed were expensive, difficult to organize, and therefore infrequent; such expeditions occurred in the 1960s and 1970s. International expeditions are still important in polar regions. Because of the enormity of poorly known ocean areas and the difficulties of working in these regions, expeditions continue to depend on international cooperation. The need for more observations has led to large, international expeditions involving ships, aircraft, and satellites of many nations working together. But we still know less about the ocean bottom, for instance, than we know about the surfaces of the Moon, Mars, and Venus.

Ships move slowly (about 400 km or 250 mi per day) and can sample or observe only the ocean near them. To study ocean processes using such limited data, it was necessary to combine observations made decades apart and often separated by thousands of kilometers. The picture of the ocean that emerged from such sparse data was an ocean that changed little in time. If we had only such limited data for the land, we could determine the climate in different locations but could say little or nothing about how weather changes over a few hours. The major gap in our knowledge about the ocean is that we know more about ocean climate than we know about ocean weather.

Studies of the ocean and its short-term interactions with the atmosphere now use many different platforms and instruments. For instance, studying ocean weather requires that the entire ocean be observed within a few days, which cannot be done using slow-moving ships. Therefore, Earth-orbiting **satellites** (Figure 1.10) now observe ocean processes. These platforms carry instruments that measure ocean-surface features such as temperature, roughness, and color. Data from these instruments are relayed back to computers, which combine the vast quantity of data into maps of the sea surface that are revised every few days. Thus we now have maps of the ocean surface comparable to weather maps of the atmosphere. In many areas, fishermen receive daily reports on current locations and water temperatures. More frequent observations from satellites will permit more accurate descriptions and therefore better predictions.

One drawback of most instruments carried on satellites is that they can sense conditions only at the ocean surface. Other techniques are therefore necessary to study currents and other oceanic processes occurring in the deep ocean. Such new techniques are being developed. (We discuss these techniques in Chapter 5.)

Submersibles, such as *Alvin* (Figure 1.11), permit detailed studies of the ocean bottom and the processes and animals that occur only there. Furthermore, they allow instruments to be carefully emplaced on the ocean bottom and recovered later. Their principal problem is their low speed and limited range. The

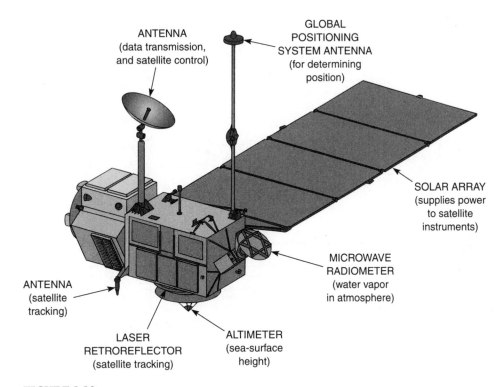

FIGURE 1.10
The ocean-observing satellite TOPEX, built by the United States and France, was launched in 1992. Data from sensors on this satellite provide nearly continuous global views of ocean conditions. This is an example of international cooperation to provide global observations of the ocean.
(Courtesy NASA)

precise area to be studied must first be located, and then manned submersibles can be used. The scientists who discovered the *Titanic,* which sank in 1912 with the loss of 1500 lives, used large-scale surveying techniques and remotely operated cameras deployed from surface ships. After the wreck had been located by remotely operated cameras, manned submersibles were used by U.S. and French scientists to survey and photograph it in detail.

Unmanned vehicles offer increased opportunities for remotely controlled ocean-bottom studies. While using these instruments, the scientist stays aboard a surface ship or may even remain in a laboratory ashore, communicating with the instruments and remotely controlling their activities. These vehicles can operate on the ocean bottom for long periods of time, and their greater depth range makes them attractive for many studies too deep for the present generation of manned submersibles.

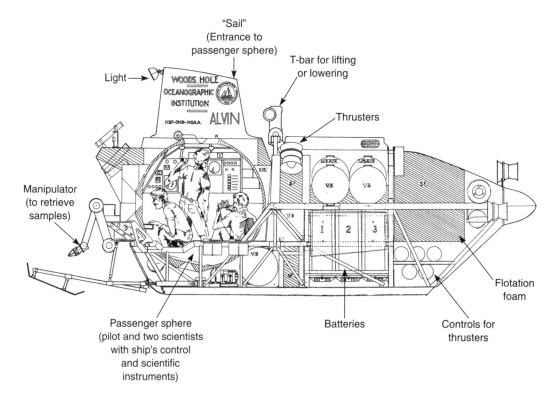

FIGURE 1.11
The submersible *Alvin* is used for detailed studies of ocean features and for placement
(and later retrieval) of instruments for studying selected processes and features. In 1986,
Alvin explored the wreck of the *Titanic* (which sank in 1912), after it was discovered in
1985 by a U.S.–French expedition. The 1986 *Alvin* dives also used a remotely controlled
vehicle, called JASON, JR., that was tethered to *Alvin*, from which it received its power
and control signals.
(Courtesy Woods Hole Oceanographic Institution)

1.8   RESOURCES

A **resource** is a supply of a desirable material, such as food, water, oil, or iron
ore. Renewable resources are replenished through growth or other processes at
rates that equal or exceed rates of consumption. Nonrenewable resources are
either not replenished or are replenished much more slowly than they are consumed. Fresh water, forests, and foods are examples of renewable resources;
petroleum and metals such as copper are nonrenewable.

The ocean provides several renewable resources, the principal one being
fresh water. Food from the sea is also a renewable resource if properly managed
(which, as we shall see, rarely happens). Ocean resources (Figure 1.12) are

important because of the depletion of nonrenewable resources on land. For example, after new supplies on land became harder to find after the 1940s, continental shelves off Texas and Louisiana were drilled for oil and gas. Exploration for oil and gas continues in these areas, especially in deeper waters farther offshore. Large petroleum reserves doubtlessly exist on continental shelves, including the little known but politically contested shelf areas off Southeast Asia.

Estuaries and continental shelves have been dredged to obtain sand and gravel after land deposits were exhausted. In many cases the coastal ocean bottom now offers the only convenient supplies of these materials, especially for small islands.

1.9   LAW OF THE SEA

As new resources (especially oil and gas) were found farther from shore, maritime political boundaries became more important. In 1982 the United Nations Conference on the **Law of the Sea** completed a treaty to clarify the legal status of these boundaries and the underlying resources. After more negotiations, the Treaty finally entered into force in November 1994.

Under the Law of the Sea, each coastal state can claim an **Exclusive Economic Zone** (EEZ) extending 200 nautical mi (370 km) seaward from their shore, including offshore islands (Figure 1.13). Large ocean areas—about 32%— and virtually all coastal oceans now are under the jurisdiction of a coastal state. Many marginal ocean basins—for instance, the Gulf of Mexico, the Caribbean Sea, and the North Sea—are totally within the EEZs of countries surrounding the basins. Several straits (narrow ocean passages connecting adjoining ocean areas) are also controlled by coastal states.

Under the Law of the Sea Treaty, many states acquired control of large ocean areas (Figure 1.13). The United States and Australia acquired the most. Their EEZs nearly equal the land areas of each country. Island states acquired control of large ocean areas around them. For instance, New Zealand's EEZ is 18 times larger than its land area, and Indonesia claims control of the waters surrounding its islands.

Designation of Exclusive Economic Zones raised many questions and caused many disputes over **maritime boundaries.** The United States has disputed its maritime boundaries with Canada over Georges Bank in the Atlantic (decided by the World Court) and at the entrance to the Strait of Juan de Fuca in the Pacific (still unresolved). The United States negotiated settlements with Russia over the Alaska-Siberia boundary, arising from uncertainties in maps used originally to define the Alaska purchase in 1867.

Some disputes involve **fisheries,** especially the several disputes between the United States and Canada. Other boundary disputes involve oil and gas. Most disputes are settled through negotiations between the countries involved. A few have been resolved in international courts. Lacking strong economic

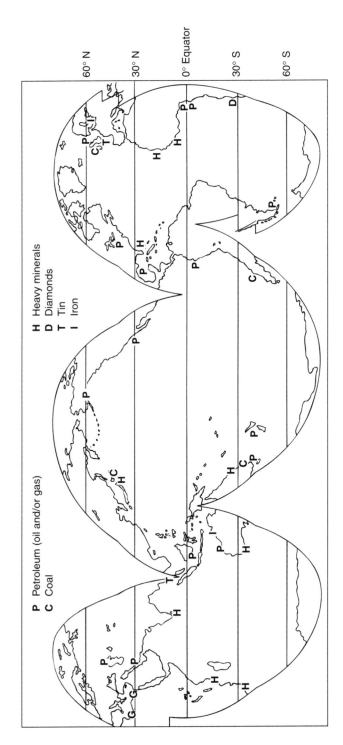

FIGURE 1.12

Fossil fuels and some industrial materials are produced from the ocean floor, mostly from continental shelves.

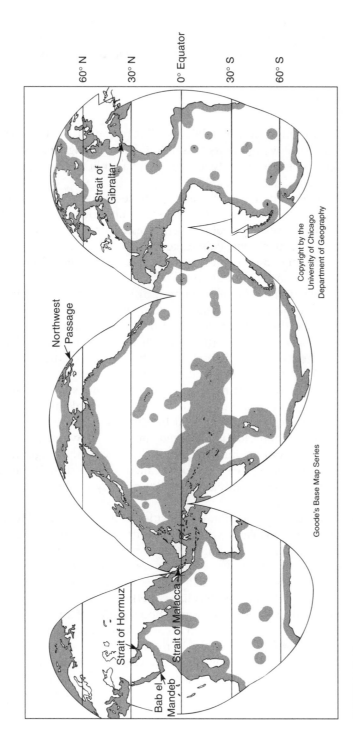

FIGURE 1.13

Exclusive Economic Zones (EEZs) of coastal states extend 200 nautical miles (370 km) from their shorelines. These EEZs cover about one-third of the ocean. Several important straits (narrow waterways connecting bodies of water), essential for both commercial and military vessels, are affected by the new boundaries and by international laws governing usage of these straits.

17

incentives, many boundary disputes will likely remain unsolved for years. The legal statuses of some straits through which ships must pass remain unresolved, such as the Northwest Passage through the Canadian Arctic islands.

1.10   NAVIGATION

Navigation (determining one's location) is vital to modern ocean exploration and research. Recent developments in satellite-based navigation systems have greatly improved navigation worldwide and facilitated both research and monitoring activities (Figure 1.14).

Early sailors remained within sight of shore, usually going ashore each night to sleep. Thus, they navigated from one landmark to another. But once having gone farther to sea, they developed celestial navigation, using the Sun and stars to locate their positions. Polynesian navigators navigated long distances across the Pacific using stars as well as wave patterns (created by the steady Trade Winds) to locate their destinations.

An early navigational scheme used by Europeans in their "Voyages of Discovery" involved the North Star, which is directly overhead at the North Pole and on the horizon at the equator. A navigator thus determined the ship's **latitude** (distance north or south of the equator) by measuring the North Star's elevation above the horizon. Once the ship had sailed to the latitude of its intended destination, it sailed either directly east or west until it reached its destination. Distances were estimated by using the ship's speed through the water and then calculating the distance the ship traveled during a day. This scheme, called dead reckoning, was used by Columbus on his exploring voyages. Problems arose when clouds obscured the stars or when the ship rolled too much to permit accurate measurements. This way of navigating did not provide measures of distances east or west (called **longitude**). The inability to determine longitude greatly limited the ability of early explorers to determine their positions east or west.

A major advance in navigation came with the invention in 1736 of clocks, called **chronometers,** capable of keeping accurate time aboard rolling, pitching ships at sea. By determining the time of local noon (when the Sun is directly overhead) one can calculate distance east or west from the Prime Meridian or any other reference point. Because Earth rotates 15° per hour, if local noon were two hours later than at the Prime Meridian, one knew that the ship was located 30° west of London. If local noon were two hours earlier than London, the ship would be at 30° east. The ability to keep accurate time at sea made it possible to determine east-west positions anywhere in the world.

The latest navigational scheme (called the **Global Positioning System,** or GPS) was completed in the early 1990s; its 24 satellites permit accurate navigation anywhere in the world (Figure 1.14). Small receivers on ships or aircraft communicate with four or more satellites. Using time and distance signals received from these satellites, the receivers determine their positions, including

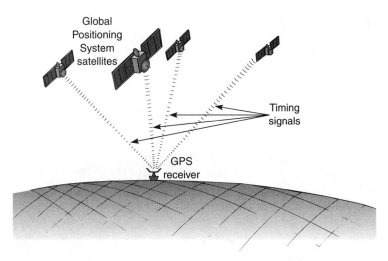

FIGURE 1.14

Schematic representation of the satellite-based Global Positioning System (GPS), used for accurate positioning of ships, aircraft, and land-based stations. Signals from four satellites are used to compute the position of the receiver, including its altitude, within a few tens of meters.

altitude. This permits ships and aircraft to navigate precisely anywhere on or over the ocean, determining their positions to within the width of a city street. This advance has revolutionized navigation and many areas of ocean research.

Now oceanographers can accurately determine their ships' movements or the positions of floating instruments, which has greatly improved measurements of currents, for example. GPS also permits recovery of instruments lost overboard.

System accuracy can be further improved by installing additional transmitters at precisely known locations. Then the receivers can locate points to within a fraction of an inch (fraction of a centimeter), also revolutionizing land surveying. Such highly accurate systems can measure land-surface displacements following earthquakes or measure slow movements of crustal plates (discussed in the next chapter).

1.11   COMMUNICATIONS

Satellite-based communications systems are essential for modern ocean studies (Figure 1.15). Research vessels now provide labs and living spaces for scientists at sea. Ocean studies are often conducted in remote locations chosen because particular processes or organisms occur there. A few decades ago, research ships had only rudimentary communications capabilities. Now, using **communications satellites,** oceanographers at sea can communicate with labs or computers ashore, sending data to be shared with their collaborators anywhere in the world.

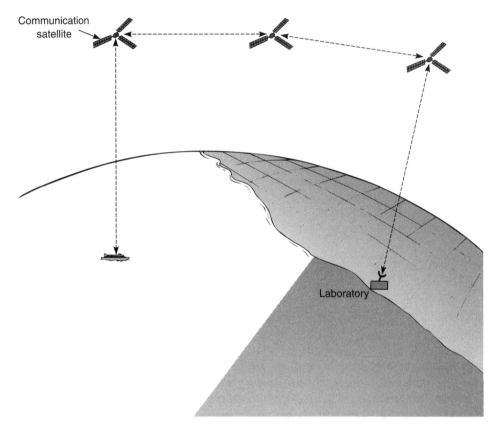

FIGURE 1.15
A satellite-based communications system, linking scientists and their data on a research
vessel with their computers and instruments ashore.

Aircraft are also used to map ocean features around the ship to assist
experimenters. Remotely sensed images of the ocean surface are also processed
and used to plan experiments, sampling, and the ship's movements while the
ship is still at sea. This too requires reliable communications.

In short, improved communications has transformed research vessels from
remote outposts collecting materials and making observations to the centers of
complex process studies drawing on people and resources around the world.

QUESTIONS

1. Explain the importance of the ocean currents around Antarctica.
2. List the important geographic feature(s) of each ocean basin. How does each feature influence the world ocean?
3. In which ocean basin are marginal seas most numerous? Least numerous?
4. Explain why the expansion or retreat of continental glaciers on land causes sea level to change.
5. Discuss the evidence for the antiquity of the ocean.
6. List and briefly discuss the processes that cause shoreline positions to change.
7. Which ocean basin receives the largest volume of river discharges? Which receives the least, relative to its volume?
8. Discuss the economic benefits to be realized from detailed, accurate forecasting of ocean currents.
9. What is the principal contribution of satellite remote sensing to the study of the ocean?

SUPPLEMENTARY READINGS

Books

Baker, D. J. *Planet Earth: The View from Space*. Cambridge, MA: Harvard University Press, 1990. Introduction to remote sensing of Earth and its ocean.

Ballard, R. D. *Exploring Our Living Planet*. Washington, DC: National Geographic Society, 1983. Elementary, well illustrated.

Ebbinghausen, E. G., and Zimmerman, R. L. *Astronomy*. 6th ed. New York: Macmillan Publishing Company, 1992. 196 pp. Elementary.

Linklater, E. *The Voyage of the Challenger*. London: John Murray, 1972. Account of the first modern ocean-exploring expedition.

Lane, N. G. *Life of the Past*. 3d ed. New York: Macmillan Publishing Company. 334 pp. Elementary. Assumes knowledge of geology.

Menard, H. W. *Anatomy of an Expedition*. New York: McGraw-Hill Book Company, 1969. Description of an oceanographic expedition.

Articles

Bailey, H. S., Jr. "The Voyage of the Challenger." *Scientific American* 188(5):88–94.

Barghorn, E. S. "The Oldest Fossils." *Scientific American* 244(5):30–54.

Browning, M. A. "Stick Charting." *Sea Frontiers* 19(1):34–44.

Bullard, Edward. "The Origin of the Oceans." *Scientific American* 221(3):66–75.

Kasting, James F.; Toon, Owen B.; and Pollack, James B. "How Climate Evolved on the Terrestrial Planets." *Scientific American* 258(2):90–98.

KEY TERMS AND CONCEPTS

Origin of universe
Big Bang
Origin of Solar System
Terrestrial planets
Asteroids
Comets
Meteorites

Origin of ocean and atmosphere
Distribution of land and ocean basins
Ocean basins
Continents
World ocean
Pacific ocean
Atlantic Ocean

Indian Ocean
Southern Ocean
Arctic Ocean
Mediterranean Sea
Surface elevations and depressions
Granitic rocks
Basaltic rocks
Challenger Expedition
Remote sensing instruments on satellites
Submersibles
Resources

Law of the Sea
Exclusive Economic Zones (EEZs)
Maritime boundaries
Fisheries
Oil and gas
Navigation
Latitude
Longitude
Chronometers
Global Positioning System (GPS)
Communications satellites

2
Dynamic Earth

In Chapter 1, we saw that space exploration has revealed Earth's uniqueness in the Solar System. Its ocean and the unusual composition of its oxygen-rich atmosphere set Earth apart from its neighboring planets. In this chapter we learn why Earth is unique, about the processes causing its uniqueness, and how these processes affect Earth's atmosphere, ocean basins, and continents.

The planet's interior is still cooling, losing heat that came originally from several sources: from Earth's formation, from later differentiation of its core, and from continued radioactive decay of materials deep in its interior. Earth cools primarily through **convection** (vertical currents caused by density differences). Rocks melted by heat from Earth's interior rise toward the surface, where they erupt as volcanoes; some heat escapes through conduction over the entire surface of the planet. After the rocks solidify and cool, they sink back toward the interior.

These convection currents cause Earth's rigid exterior plates to move, thereby moving continents as well as forming, and later closing, ocean basins. Collectively, this process is called **plate tectonics.** In this chapter we discuss these processes and how they have shaped and reshaped Earth over its more than four-billion-year history.

In passing, it is worth noting that Mars also underwent such cooling processes in the past. But being smaller than Earth, it apparently cooled so much that convection ceased and is essentially dead. The Moon and Mercury, much smaller than Earth, apparently did not undergo these processes, because they cooled too quickly. (Whether Venus is still undergoing tectonic activity is disputed.)

We begin by examining Earth's internal structure and then proceed to consider how its continued cooling works and how it has affected Earth and its ocean.

2.1 STRUCTURE OF EARTH

The nearly spherical Earth has two outer fluid layers: ocean and atmosphere. Earth's interior (Figure 2.1) consists of nearly **concentric shells:** the least dense, rigid crust at Earth's surface, the denser and more mobile mantle, and the extremely dense core at Earth's center. Each shell has a different composition and density. Most of our evidence about Earth's interior comes indirectly from studies of earthquake waves whose directions and speeds change as they pass through materials having different characteristics.

Earth's density is determined from studying how Earth affects other bodies, such as the Moon. Much of our knowledge of Earth's interior composition comes from studying meteorites, which apparently come from broken-up planetlike bodies, orbiting the Sun between Mars and Jupiter.

Earth's **core** contains its densest materials, consisting of a solid iron-nickel alloy (inner core) and a liquid iron-nickel outer core. Because of its great density, the core constitutes about 31% of Earth's mass. The liquid outer core circulates, generating Earth's magnetic field; this magnetic field changes (reverses) at irregular intervals over geologic time. These magnetic reversals provide evidence that scientists can use to determine crustal ages, to study ocean-basin history, and to reconstruct plate movements. (There will be more about this in later sections of this chapter.)

The **mantle,** consisting of relatively dense rocks, makes up about two-thirds of Earth's mass. The lower mantle is poorly known but appears to be relatively homogeneous. We still do not know if the lower mantle participates actively in Earth's internal circulation. The upper mantle is clearly involved.

FIGURE 2.1
Schematic representation of Earth's concentric, internal structure. The core contains the densest materials. The crust contains the least dense rocks. The ocean and atmosphere are not shown.

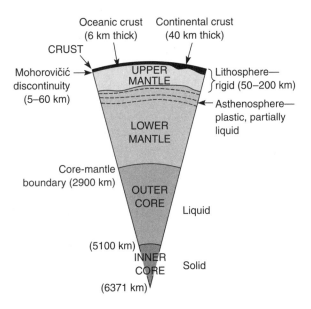

The outermost mantle is relatively rigid and moves with the overlying rigid crust. The mantle immediately below this rigid layer, called the **asthenosphere,** is easily deformed, relatively weak, and probably near its melting point. The outer layers of the mantle and the crust, together called the **lithosphere,** float in the asthenosphere and also move across it.

Earth's outer layer, called the **crust,** consists of two different types of rocks: **continental crust** (averaging 40 km, or 25 mi, thick) and **oceanic crust** (averaging 7 km, or 4 mi, thick). Both the crust and the outer mantle rocks are cool and relatively rigid and move together. (There will be more about these rigid plates in the following sections.)

The lithosphere is thickest under continents (several hundred kilometers) and thinnest under the ocean basins, where it may be as little as a few tens of kilometers thick under the mid-ocean ridges. As the lithosphere cools, it thickens, becomes denser, and floats lower in the asthenosphere.

2.2   PLATE TECTONICS

The large, rigid lithospheric plates (Figure 2.2) move as units, driven by movements of underlying mantle. This process (**plate tectonics**) causes many of Earth's unique features.

New lithosphere is formed at **mid-ocean ridges** by intermittent volcanic eruptions, and the oldest lithosphere is absorbed into the mantle at **trenches.** Plate movements are slow, typically 2 to 3 cm (about 1 in.) per year (about the rate your fingernails grow). In the South Pacific, plate movements are faster, up to 20 cm (8 in.) a year. These slow plate movements eventually cause the breakup of continents and the formation of ocean basins.

There are seven large plates (each larger than continents) and many smaller ones (Figure 2.2). Most plates include parts of continents as well as ocean basins. There are three different types of **plate boundaries:** (1) mid-ocean ridges, where plates form; (2) trenches, where plates are destroyed; and (3) **transform faults,** large fracture zones where plates slide past each other (Figure 2.3).

Plate tectonics explains many of Earth's enigmatic surface features, among them the close fit between the opposite shores of the Atlantic Ocean. About 200 million years ago, a large continent (called **Pangea**) broke up, and the pieces moved to their present locations.

Lithospheric plates are rigid, so there is little faulting (cracking of the crust) or volcanic activity, and no mountain building occurs within plates. These processes occur almost exclusively at plate edges, which are marked by earthquakes and active volcanoes. In fact, earthquakes are used to map plate boundaries in little-known areas. Earthquakes can be detected and their locations determined at great distances by networks of **seismometers** (instruments that detect and record earthquake waves).

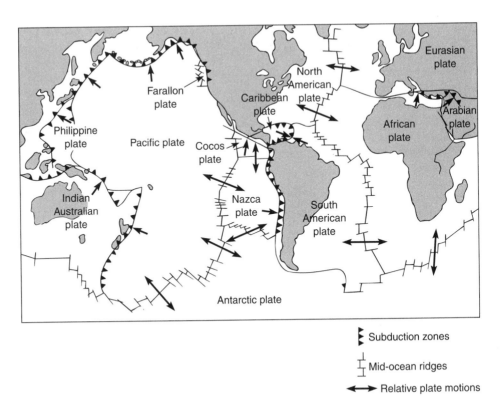

FIGURE 2.2
Map of major crustal plates, which are bounded by mid-ocean ridges (divergence zones
where plates are formed), by trenches (convergent zones where plates are destroyed),
and by transform faults (where plates slide past each other). Frequent earthquakes occur
on plate boundaries. Active volcanoes occur only near trenches.

2.3   MID-OCEAN RIDGES

New lithosphere forms at mid-ocean ridges by volcanic eruptions (Figure 2.3).
Fissures (cracks in the crust) open up as plates move away from ridge axes. Such
fissures can be observed on Iceland, which is an exposed part of the Mid-Atlantic
Ridge. Shallow earthquakes at ridge crests are often associated with volcanic
eruptions. Eruptions of Hawaiian volcanoes are easily observed and provide
much of our knowledge about mid-ocean ridge processes, which have been
mostly hidden from study. Now scientists can detect active seafloor volcanoes,
using the sounds emitted during earthquakes and volcanic eruptions. Knowing
where eruptions are occurring, it is now possible to send ships and submersibles
and to place instruments to study events as they happen, rather than having to
reconstruct them from evidence gathered years later.

New crust forms at mid-ocean ridges by intermittent eruptions of large volcanoes, much like those on Hawaii. As hot lava (at about 1200°C, or 2200°F)

contacts cold seawater (at 2 to 5°C, or 36 to 41°F), its outer surface hardens, forming tubelike and pillow-shaped structures (Figure 2.4). Larger volcanic eruptions can form sheet flows that have chilled upper surfaces under which molten lava can still flow.

Where crustal formation is relatively slow (low spreading rates are typically less than 6 cm, or 2.4 in., per year), mid-ocean ridges have rugged rift valleys, 1 to 2 km (0.6 to 1.2 mi) deep and a few tens of kilometers wide; the slowly spreading Mid-Atlantic Ridge is an example. More rapid spreading results in less rugged, broader domelike rises, usually with no rift valleys; this is typical of the rapidly spreading East Pacific Rise.

Heat is removed from newly formed crustal rocks by two processes. Conduction upward through the ocean floor removes about one-third of the heat flowing through mid-ocean ridges. The rest is removed by cold seawaters circulating through rocks on the sea floor. These waters are heated as they flow through cracks and fissures in the ocean bottom, whereas the hottest waters penetrate many kilometers below the sea bottom, removing heat from still cooling volcanic rocks. The heated waters rise to the ocean bottom and discharge through spectacular vents, releasing especially hot waters (400°C, or 750°F). This set of processes, called **hydrothermal circulation,** apparently also controls sea-salt composition (discussed further in Chapter 3).

Where spreading rates are rapid (more than 6 cm, or 2.4 in., per year), much of the superheated water discharges through vents. Minerals precipitate out of these hot waters as they mix with cold, oxygen-containing ocean waters and form conical structures up to 10 m (33 ft) high (Figure 2.5). Copper and zinc deposits formed in this manner are mined on land, but not from deep-ocean deposits.

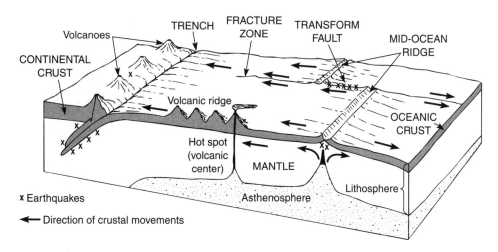

FIGURE 2.3
Crustal plates and their movements. New oceanic crust is formed by mid-ocean ridges and destroyed at trenches.

FIGURE 2.4
Lava erupted from sea-floor volcanoes forms tubelike and pillow-shaped structures. This
photograph was taken from the submersible *Alvin*.
(Courtesy Woods Hole Oceanographic Institution)

2.4   OCEAN-FLOOR AGES AND DEPTHS

Magnetic patterns recorded by rocks on the sea floor can be used to decipher
plate movements (Figure 2.6). Let's see how the technique works. As lavas cool,
newly formed iron-containing magnetic minerals record the orientation of
Earth's magnetic field at that time. Patterns of **magnetic anomalies** (stripes of
alternately stronger and weaker magnetic intensity) resulting from the magnet-
ism retained by these minerals can be determined by airborne or shipborne
instruments called magnetometers. Rocks or sediments from the ocean bottom
can also be used to determine orientations of magnetic minerals in them.

During Earth's **magnetic-field reversals,** its north and south magnetic
poles exchange places. As new oceanic crust solidifies at mid-ocean ridges, its
iron-containing minerals record the orientation of Earth's magnetic field at that
time. Thus the deep-ocean floor can be viewed as a gigantic tape recorder,
recording Earth's changeable magnetic field over the last 200 million years.

Rocks formed when Earth's magnetic field was oriented the same as it is
now contribute to the strength of the present field, forming a positive anomaly.

FIGURE 2.5
Hot waters (350°C, 650°F) discharged by a chimneylike structure, called a "black smoker," on the East Pacific Rise. Sulfides in vent waters react chemically with seawaters, precipitating metal-sulfide particles, which form smoker chimneys and also accumulate in nearby sediment deposits.
(Photograph by Robert D. Ballard. Courtesy Woods Hole Oceanographic Institution)

If the rocks were formed at times when the magnetic field was reversed, the magnetic field above them is slightly weaker than the average field. The combination of stripes of slightly stronger and slightly weaker fields forms the pattern of magnetic anomalies used by scientists to determine when a particular segment of oceanic crust formed.

Volcanic rocks on land also record these magnetic reversals. By combining many such studies, a time scale of reversals has been established. From patterns of magnetic anomalies in ocean basins, scientists can determine when ocean basins formed and how continents have moved over the most recent 200 million years. Earlier reconstructions are less certain because the data are not as good.

FIGURE 2.6
Magnetic anomalies on the ocean floor are generated by magnetic minerals in ocean-bottom rocks, which recorded the orientation of Earth's magnetic field when the minerals formed.

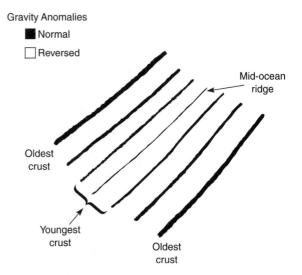

Magnetic anomaly patterns show that much of the Atlantic oceanic crust formed during the past 80 million years. The oldest crust formed about 160 million years ago (Figure 2.7). These magnetic patterns also provide details on ocean basin formation. For example, we know that the North Atlantic began opening before the South Atlantic, because the oldest North Atlantic crust is 160 million years old, whereas the oldest crust in the South Atlantic is only 140 million years old (Figure 2.7). (The Caribbean Plate between North and South America has had a separate history, apparently forming in the Pacific about 195 million years ago and moving into its present position much later.)

Newly formed ocean floor at mid-ocean ridges stands higher than its surroundings because warm crustal rocks are less dense than older and colder rocks. As oceanic crust ages and moves away from mid-ocean ridges, it cools, becomes denser, and sinks lower into the asthenosphere. In other words, the older the crust, the deeper it lies below the ocean surface (Figure 2.8). The ocean floor not only acts as a tape recorder for magnetic reversals, but also resembles a gigantic conveyor belt that deepens as it ages. At mid-ocean ridges, the ocean bottom is about 2.5 km (1.5 mi) deep, but is about 6 km (3.7 mi) deep when it is 160 million years old (Figure 2.8). The oldest known oceanic crust in the western North Pacific has been sampled by **scientific drilling** operations.

2.5 TRENCHES AND ISLAND ARCS

Lithosphere is absorbed at trenches (Figure 2.3) where plates are pulled down into the mantle, a process called **subduction.** Both oceanic crust and any sediments deposited on it sink into the mantle, eventually melt, and mix with other mantle materials. Melting of lithosphere and sediment in the mantle supplies magma (molten rock) to volcanoes situated above subduction zones because plumes of lighter molten rock rise toward the surface and erupt, forming volcanoes (Figure 2.9).

Where one oceanic plate converges with another, volcanic activity caused by subduction forms island arcs. Trenches usually lie on the seaward sides of island chains with relatively shallow seas near continents.

Trenches and island arcs surround most of the Pacific, especially the western Pacific. The only gaps in the circum-Pacific trench/island-arc system (called the Pacific Rim of Fire) are in Antarctica and in western North America.

Deep earthquakes (at depths greater than 100 km, or 60 mi) are characteristic of subduction zones. Earthquakes occur as adjacent crustal blocks drag past each other (Figure 2.9). These earthquakes release energy stored in rocks deformed during plate movements. The deepest earthquakes occur at depths of 700 km (430 mi). On mid-ocean ridges, earthquakes occur at relatively shallow depths, from a few kilometers to a few tens of kilometers deep.

FIGURE 2.7
Age of oceanic crust of the Atlantic Ocean, based on patterns of magnetic-field reversals in ocean-floor rocks. Note that the North Atlantic basin began forming earlier than the South Atlantic. The history of the Caribbean Plate is still unknown, but it apparently formed in the Pacific and moved to its present location between North and South America. Its oldest parts are 30 million years older than the oldest parts of the North Atlantic.

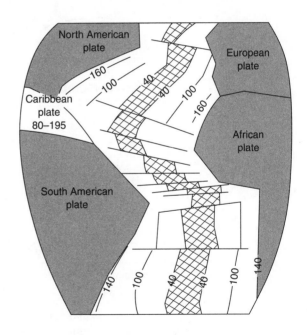

FIGURE 2.8
Ocean-floor depth increases with crustal age.
[After J. G. Sclater, R. N. Anderson, and M. L. Bell. "The Elevation of Ridges and Evolution of Eastern Pacific," *Journal of Geophysical Research,* 76 (1971), pp. 7888–7915]

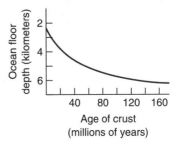

(a) OCEAN-OCEAN CONVERGENCE

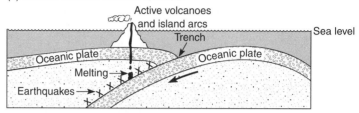

(b) OCEAN-CONTINENT CONVERGENCE

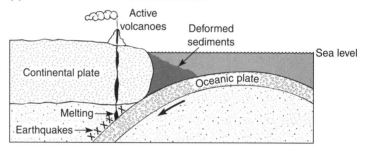

(c) CONTINENT-CONTINENT COLLISIONS

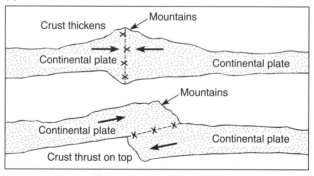

X Earthquakes ◀── Directions of plate movements

FIGURE 2.9
Schematic representation of different types of plate convergences. (a) Ocean-plate convergences occur in the Aleutian Islands of Alaska and in the Japanese islands. Conspicuous trenches occur in both areas. (b) Oceanic plates are being subducted beneath continental plates in North and South America. Sediments on the ocean floor are deformed by the collision. Trenches occur only where there are few sediments. (c) Continental plates converge in India and in Turkey. The crust on both plates is thickened by the collision, forming mountain ranges.

Continents can also be pushed together, but such convergences do not form trenches or island arcs. The most conspicuous continental convergence is in the Mediterranean region, where Africa and Europe are colliding. There is a small trench, the Hellenic Trench, south of Greece, but no obvious island arc, although there are active volcanoes in Italy and Greece. In Turkey and Pakistan, the continuing collision of Asia and Africa is marked by destructive earthquakes and mountain building.

During the last 40 million years, the Indian subcontinent has moved thousands of kilometers northward and collided with the Asian continent. The most conspicuous result is the Himalayan Mountains, relatively young mountains that were formed where part of the Indian block was thrust beneath the Asian block. The Rocky Mountains of western North America were formed when the Pacific plate thrust beneath North America; their present height results from the buoyant, relatively young Pacific plate materials lying beneath the North American continental crust. In other words, the Rocky Mountains are underlain by doubly thick crust.

2.6   FRACTURE ZONES AND TRANSFORM FAULTS

Lithospheric plates slip past each other along transform faults. Earthquakes or active volcanoes mark the active parts of transform faults (Figure 2.3). Inactive parts of transform faults are marked by fracture zones (linear bands of rugged ocean bottom topography), but no earthquakes or active volcanoes.

Earthquakes occur only on active transform-fault segments, between active ridge crests, where plates move in opposite directions on each side fault. Beyond the active ridge-crest segments, plates on both sides of transform faults move in the same direction and at the same speeds; thus, there are no earthquakes. Traces of transform faults persist for millions of years until they are buried by sediment deposits. These fault traces record ancient plate movements. Fracture zones are conspicuous in the central Pacific, which receives too little sediment to bury them. (Reasons for the lack of sediment in the central North Pacific will be discussed in Chapter 9.)

2.7   CONTINENTAL MARGINS

Continental margins mark transitions between two different kinds of crustal rocks: less dense, granitic, continental crust and more dense, basaltic, oceanic crust. Continental margins generally do not coincide with plate boundaries. They are formed by the same plate movements that shape ocean basins. There are three major types of continental margins, distributed as shown in Figure 2.10.

Divergent margins (also called **passive** or **Atlantic-type margins**) form when continents are fractured and pulled apart, forming new ocean basins (Figure 2.11). These continental margins mark transitions between oceanic crust

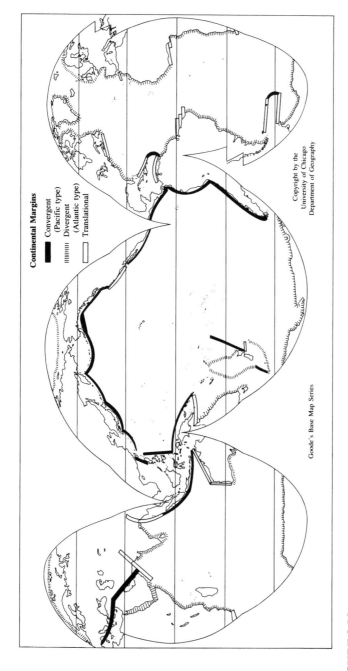

Continental Margins

- ▬ Convergent (Pacific type)
- ‖‖‖ Divergent (Atlantic type)
- ▭ Translational

Goode's Base Map Series

Copyright by the
University of Chicago
Department of Geography

FIGURE 2.10

Distribution of different types of continental margins. Note that the Pacific basin is nearly surrounded by trenches (Pacific-type margins) The Atlantic-type margins are mostly divergent margins, except where subduction is occurring in the South Sandwich Trench (south of South America) and in the West Indies, between North and South America. [After K. O. Emery, "Continental Margins—Classification and Petroleum Prospects," *Bulletin of the American Association of Petroleum Geologists* 64 (1980), pp. 297–315] and continental crust. The passive continental margins around the Atlantic are conspicuous examples.

Divergent margins form by thinning of oceanic crust or faulting of continental crust at mid-ocean ridges. Through time, these fractured continental pieces move apart and gradually subside as they cool. Because of their subsidence, these continental margins accumulate thick deposits of sediments eroded from nearby lands.

Convergent margins (also called **active** or **Pacific-type margins**) mark the boundaries between converging crustal plates (Figure 2.12). These are some-

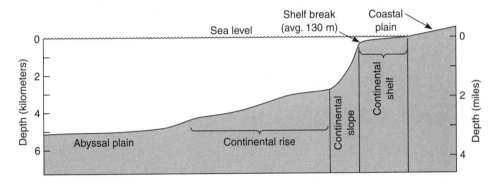

FIGURE 2.11
Profile of a continental margin (Atlantic-type), also called a passive margin. Note that the steepness of the slopes is greatly exaggerated in these profiles.

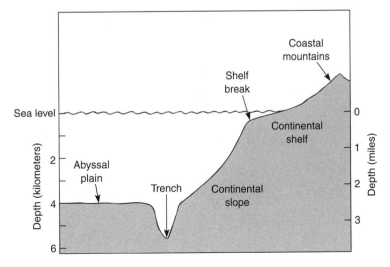

FIGURE 2.12
Schematic profile of a Pacific-type margin, also called an active margin.

times associated with subduction. They can thrust an oceanic plate under a continental plate, which is now occurring along the Pacific coast of North and South America. Or a continental block can be thrusting under another, as in the Himalayan Mountains of northern India.

Translational margins (also called **transform margins**) are formed by lateral motions between plates. They may be marked by shallow earthquakes. During rifting, parts of continental crust move relative to the adjacent crustal plate. At present, parts of California (west of the San Andreas Fault) and Baja California, Mexico are moving northwest relative to the North American plate. Eventually parts of Southern California and Baja California will form a large island, offshore from North America. The Channel Islands, off Southern California, are pieces broken off the continent many million years ago, probably by the East Pacific Rise.

2.8 VOLCANOES AND VOLCANIC ISLANDS

Most submarine mountains are extinct volcanoes, as are many oceanic islands. Some ancient, extinct volcanoes are now submerged and covered by coral reefs and deposits of carbonate sediments.

Where molten rock erupts at Earth's surface, lava and ash formed during the eruptions accumulate, forming mountains. About 20 major volcanic eruptions occur each year on land. Many more eruptions occur on the ocean bottom, usually unobserved. Only shallow submarine eruptions (Figure 2.13) and those occurring on islands have been extensively studied. As we have learned, this is changing rapidly, because of access to more sensitive techniques than those previously available to civilian scientists.

Unless renewed by continued large eruptions, volcanic islands are quickly eroded down to sea level by weathering and wave action. Thus, intermittent, small volcanic eruptions often form rapidly disappearing islands. These volcanic cones built on submerged banks are quickly worn down by waves or destroyed by explosions associated with later eruptions. As we discussed previously, volcanoes are also submerged as the plates on which they formed subside over millions of years (Figure 2.14). Thus, most volcanoes far from the mid-ocean ridge that formed them are usually submerged seamounts. As we will learn in Chapter 8, corals growing on them may sometimes keep up with the rate of sinking to maintain themselves at the surface as reefs or coral islands.

Long-lived **volcanic centers,** called **hot spots,** occur in the interiors of plates. They form lines of active volcanoes and volcanic islands as the plates move across them. Active volcanoes occur immediately above the hot spot, and as the plate continues to move, the volcanic cone is moved away from the hot spot and eventually becomes extinct. A new cone forms above the hot spot and continues to be active until it too moves too far from the source of molten rock. In this way a chain of volcanic islands or a volcanic ridge forms on the sea floor.

FIGURE 2.13
Surtsey, a young volcanic island near Iceland, as it appeared in 1970. After eruptions cease, the island will be eroded by waves and rain, until only a shallow submerged bank remains. (Courtesy Icelandic Airlines)

The most conspicuous example of islands formed by a plate moving over a hot spot is Hawaii (Figure 2.15). Active volcanoes are at the southeastern end of the chain, on the island of Hawaii. A new, still-submerged volcano called Loihi is now forming southeast of Hawaii. The oldest and most eroded islands, now marked by coral atolls, lie at the northwestern end. A chain of deeply submerged seamounts, the Emperor Seamounts, extends north-south on the Pacific Ocean bottom. This marks the trace of the plate over the Hawaiian hot spot. The change in direction marks the Pacific plate's change in direction about 40 million years ago.

Growth of volcanoes, whether on land or on the ocean bottom, is not a continuous process. Large explosions, some caused by water contacting molten rock, can blast away parts of the mountain. Krakatoa, a volcano between Java and Sumatra in Indonesia, exploded in 1883—one of the largest volcanic explosions in recorded history.

Following 200 years of inactivity, Krakatoa was first moderately active for several weeks. Then, on August 26, a violent explosion, heard 5000 km (3000 mi) away and measured around the world, began two days of explosions that obliterated two-thirds of the island. The explosions caused gigantic ocean waves called **tsunamis** (discussed in Chapter 6), which destroyed nearby low-lying villages and killed 36,000 people. Waves generated by the explosions formed gigantic breakers that swept inland, reaching as high as 40 m (130 ft) above sea level and 16 km (10 mi) inland on Java.

In addition, thick, windblown ash deposits formed over 770,000 km^2 (300,000 mi^2); thinner deposits covered 3.8 million km^2 (1.5 million mi^2). About

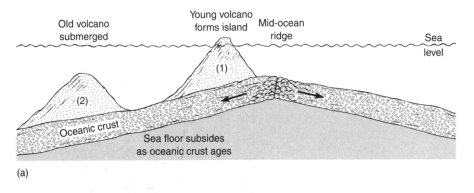

Old volcano submerged

Young volcano forms island

Mid-ocean ridge

Sea level

(1)

(2)

Oceanic crust

Sea floor subsides as oceanic crust ages

(a)

ISLAND BUILDING

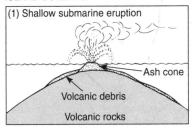

(1) Shallow submarine eruption

Sea level

Ash cone

Volcanic debris

Volcanic rocks

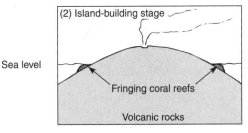

(2) Island-building stage

Fringing coral reefs

Volcanic rocks

ISLAND SUBSIDENCE AND EROSION OF VOLCANO

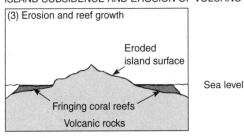

(3) Erosion and reef growth

Eroded island surface

Sea level

Fringing coral reefs

Volcanic rocks

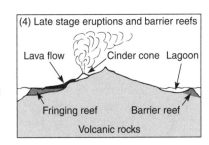

(4) Late stage eruptions and barrier reefs

Lava flow

Cinder cone

Lagoon

Fringing reef

Barrier reef

Volcanic rocks

ATOLL STAGE AFTER SUBSIDENCE

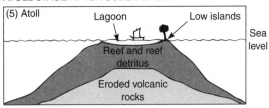

(5) Atoll

Lagoon

Low islands

Sea level

Reef and reef detritus

Eroded volcanic rocks

(b)

FIGURE 2.14

(a) Recently formed volcano (1) is eroded by winds and waves and is eventually worn down. Even if it is not completely eroded down, it gradually subsides (2) as the oceanic plate under it cools and subsides. (b) Stages in the evolution of coral reefs as the volcanic base is gradually eroded down to sea level. Coral reefs and eventually coral atolls (ring-shaped islands surrounding shallow lagoons) mark the site of the volcano. Note the subsidence of the volcanic foundation between (4) and (5), leaving only the limestone and carbonate sand islands at the surface.

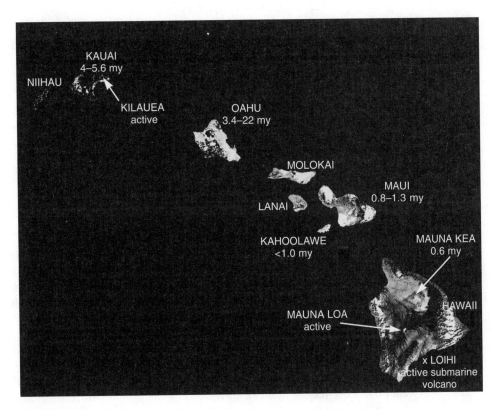

FIGURE 2.15
Hawaii, the youngest and largest of the Hawaiian Islands, lies at the eastern (youngest)
end of the chain. Hawaii has three active volcanoes: Mauna Loa, and Kilauea (on land),
and Loihi, which is still submerged on the flanks of Hawaii's southeast coast. Kauai and
Niihau, the oldest islands shown, have been eroded by waves and weathering and are
much smaller than Hawaii. The times when the volcanoes on each island were active are
given in millions of years.
(Courtesy NASA)

16 km³ (3.8 mi³) of ash was blown into the air, and deposits were 15 m (50 ft)
thick on nearby islands. Large amounts of ash were deposited on the nearby
deep-ocean floor, as well as being carried 50 km (30 mi) into the stratosphere,
where it remained for several years. During that time the ash caused brilliantly
colored sunsets throughout the Northern Hemisphere.

Another Indonesian volcano, Tambora, erupted even more massively in
1815. Its explosive eruption injected so much ash into the stratosphere, reducing
the amount of energy reaching Earth's surface from the Sun, that snow fell in New
England in July. This was called the "year with no summer." Millions starved to
death in China and probably elsewhere as well where records are not as good.

An even larger volcanic explosion on the Greek island of Santorini
around 1650 B.C.E. gave rise to the Atlantis legend. That ancient "lost city"

was destroyed by the enormous explosions and resulting gigantic waves. The widespread destruction apparently hastened the decline of the Minoan civilization in the eastern Mediterranean and led to the subsequent rise of classical Greek culture.

A large recent volcanic event occurred in 1991 when the Philippine volcano Pinatubo erupted. It injected huge amounts of ash into the stratosphere, which produced spectacularly colorful sunsets for years. The ash in the upper atmosphere also reduced the amount of sunlight reaching Earth's surface, resulting in several years of colder-than-normal winters worldwide.

2.9 SUPERCONTINENT CYCLES

Earth has been cooling since its formation. Since its crust first formed, Earth has probably undergone six, or maybe seven, cycles of supercontinent formation (when all continental pieces are brought together). **Supercontinents** later break up, forming one or more new ocean basins, which first grow and later disappear when new supercontinents form. The entire cycle from start to finish takes about 500 million years (Figure 2.16).

Our knowledge of plate movements is relatively good only for the last 500 million years. During this time, marine animals had developed the ability to form durable shells, which can be used to date the rocks containing them fairly precisely. Dating older rocks devoid of fossils requires the use of radioactive elements. These techniques are now becoming more accurate and may soon permit reconstructions of earlier supercontinent cycles. Let's now look at a single cycle, based primarily on the last 500 million years.

The cycle begins because continental crust is thick and does not conduct heat as well as oceanic crust. Thus, a supercontinent remaining in one spot for a long time causes the underlying mantle to heat because continental crust acts like a blanket, retarding heat flow from Earth's interior.

As the underlying mantle warms, it expands, elevating the overlying continent and stretching the crust. Eventually the crust breaks, forming **rift valleys,** as is now happening in East Africa. (Africa has been in its present location for about 200 million years and stands about 400 m higher than other continents as a result of thermal expansion of the underlying mantle.)

As the continental crust rifts, deep valleys form and first fill with fresh water. In East Africa, long, narrow, deep lakes now fill these rift valleys. But eventually, a long narrow ocean forms, such as the Red Sea, which is already becoming a narrow ocean basin. Eventually a mid-ocean ridge forms in the narrow ocean basin, causing it to widen.

Through time, the earliest formed lithosphere in mature ocean basins cools, eventually becoming dense enough to sink into the underlying mantle. This is now happening in the South Atlantic (Scotia Arc) and in the West Indies (near Barbados). Ocean basins typically widen for about 200 million years.

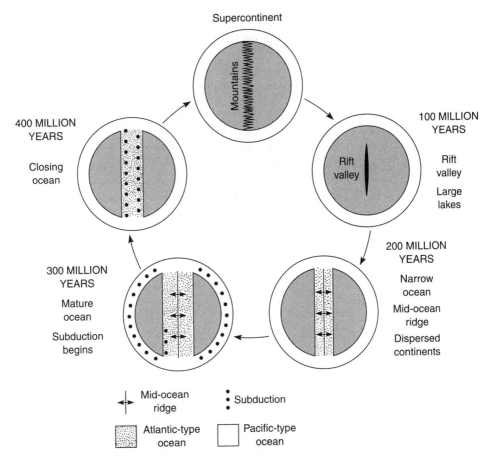

FIGURE 2.16
Supercontinents form by aggregation of continents and continental fragments. They break up to form new ocean basins, which again close to form new supercontinents. One cycle takes about 500 million years from start to finish. The same landmass is shown in all five panels.

As ocean basins age, subduction becomes more widespread around their margins. Eventually the basins begin to close as subduction overwhelms spreading at the mid-ocean ridge. For the next 200 million years, the ocean basin closes. As an Atlantic-type basin closes, its sediment deposits are deformed, forming mountain ranges on the newly assembled supercontinent that marks the site of the former ocean. The Appalachian Mountains of eastern North America formed in this way at the end of the previous supercontinent cycle, about 450 million years ago.

The Pacific Ocean basically remains intact during these cycles, although its size and shape change. As an Atlantic-type ocean widens, subduction occurs to accommodate the lithospheric blocks that are moving over its margins. Thus, the Pacific is now bordered by subduction zones, except for a few locations. When the Atlantic-type basin begins closing, most subduction in the Pacific-type-basin ceases as subduction expands in the closing basin.

The supercontinent cycle also affects **sea level** relative to the continents. During supercontinent phases, continents stand high as the mantle under them warms; as spreading begins, the continental fragments move away from the heated mantle and thus stand lower, with respect to the sea surface. Furthermore, the newly formed ocean basin floor is relatively shallow, and thus low-lying continental areas are flooded.

As the newly formed basin widens and its crust ages, cools, and deepens, sea level falls. Sea level has apparently varied by as much as 250 m (820 ft) over the past 100 million years. During times of higher seal level, ocean covers shallow areas of continents, such as Canada's Hudson Bay. At such times, oceans cover about 80% of Earth's surface. At times of lower sea level, shorelines move seaward to occur at the continental margins; the ocean then covers only about 65% of Earth's surface. These changes in the amount and distribution of shallow seas affects the distribution of marine life.

QUESTIONS

1. What fraction of Earth's surface is underlain by continental rocks?
2. List the major parts of the continental margin.
3. Describe island arcs and their origin.
4. Which type of plate boundary is characterized by deep earthquakes?
5. Briefly describe and compare the structures and compositions of ocean basins, and of continental blocks.
6. Which sea-floor features are parts of the continents? Which are parts of the ocean basins? How are they distinguished from each other?
7. Describe the processes involved in plate tectonics.
8. In the western Pacific Ocean, what sea-floor features commonly mark the boundary between ocean basin and continent?
9. What causes volcanic activity at subduction zones?
10. Discuss the relationship between crustal ages and ocean depths.

SUPPLEMENTARY READINGS

Books

Ernest, W. G. *The Dynamic Earth*. New York: Columbia Press, 1990. Elementary introduction to geology and plate tectonics.

Hsu, Kenneth J. *The Mediterranean Was a Desert*. Princeton, NJ: Princeton University Press, 1983.

Miller, Russell. *Continents in Collision*. Alexandria, VA: Time-Life Books, 1983. Well-illustrated elementary treatment of plate tectonics, including development of the theory and history of plate movements.

Seibold, E., and Berger, W. H. *The Sea Floor: an Introduction to Marine Geology*. Heidelberg, Germany: Springer-Verlag, 1993. Summary of Marine Geology.

Sullivan, Walter. *Continents in Motion: The New Earth Debate.* 2d ed. New York: American Institute of Physics, 1991. Excellent discussion of the origins of plate tectonic theory.

Articles

Dewey, J. F. "Plate Tectonics." *Scientific American* 266(5):56–68.

Emery, K. O. "Continental Shelves." *Scientific American* 221(3):106–125.

Fitfield, R. "The Structure of the Earth." *New Scientist* 1988(25 February):1–4

Heezen, B. C. "Origin of Submarine Canyons." *Scientific American* 229(5):102–112.

Heezen, B. C., and MacGregor, I. D. "The Evolution of the Pacific." *Scientific American* 251(1):46–55.

Hekinian, Roger. "Undersea Volcanoes." *Scientific American* 251(1):46–55.

Hoffman, K. A. "Ancient Magnetic Reversals." *Scientific American* 258(5):76–83.

Matthews, S. W. "This Changing Earth." *National Geographic* 143:1–37.

Menard, H. W. "The Deep-Ocean Floor." *Scientific American* 221(3):126–145.

Nance, R. D.; Worsley, T. R.; and Moody, J. B. "The Supercontinent Cycle." *Scientific American* 259(1):72–77.

Sclater, J. G., and Tapscott, C. "The History of the Atlantic." *Scientific American* 240(6):156–175.

KEY TERMS AND CONCEPTS

Earth's cooling
Convection
Earth's concentric structure
Core
Mantle
Asthenosphere
Lithosphere
Crust
Plate tectonics
Pangea
Earthquakes
Seismometers
Mid-ocean ridges
Hydrothermal circulation
Ocean-floor ages and depths
Magnetic anomalies
Magnetic-field reversals

Scientific (ocean basin) drilling
Trenches and island arcs
Subduction
Fracture zones
Transform faults
Continental margins
Divergent margins
Convergent margins
Volcanoes and volcanic islands
Coral reefs
Atolls
Volcanic centers (hot spots)
Supercontinent cycles
Supercontinents
Rift valleys and lakes
Sea-level changes

3

Seawater

Water—the most abundant substance on Earth's surface—profoundly affects the behavior of the ocean and atmosphere. To understand the importance of water and sea salts to ocean processes, we begin with the physical and chemical behavior of pure water and its properties and then examine the effects of adding salts. Finally, we look at some resources extracted from seawater.

3.1   WATER MOLECULES

Each **water molecule** consists of one oxygen atom bonded to two hydrogen atoms. Because of the oxygen atom's electron cloud configuration, the hydrogen atoms in a water molecule occupy two corners of a pyramid-shaped, four-cornered molecule (Figure 3.1). A water molecule has two hydrogen atoms on one side, and each hydrogen atom shares its single electron with the oxygen atom. Hydrogen atoms thus act as a positive charge on the water molecule. On the opposite side is the oxygen atom, which has excess negative charges because of its shared electrons. Because a water molecule's positive and negative charges are separated, it is called a **polar molecule.**

All molecules attract each other. Most attractions are weak and arise from interactions between the atomic nuclei of one molecule and the electrons of the other molecule when both are close together. In water, separation of the charges and the presence of hydrogen atoms together give rise to stronger interactions known as hydrogen bonding. **Hydrogen bonds** between water molecules are only about 1/20 as strong as the bonds between hydrogen and oxygen within the molecule itself. Nonetheless, they are strong enough to influence the properties

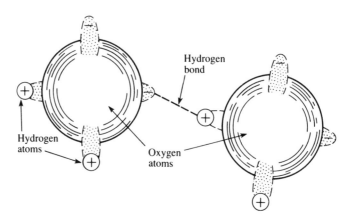

FIGURE 3.1
Each water molecule has a four-cornered structure. The two molecules are shown linked by a hydrogen bond, an interaction between a hydrogen atom of one molecule and a negative charge on the oxygen atom of the other molecule.

of liquid water. If hydrogen bonding were not present, ice would melt at about –100°C (–148°F) and water would boil at about –80°C (–112°F), and water would exist only as a gas at Earth's surface temperatures and pressures. Without hydrogen bonding of water molecules, there would be no ocean and no life on Earth.

3.2 FORMS OF WATER

Water is one of the few common substances on Earth that readily exists in all three **forms of matter:** crystalline solid (ice); liquid (water); and gas (water vapor). Let us consider each form to see how water's internal structure controls its properties in each of these three states.

Like all crystalline solids, ice has an orderly internal structure (Figure 3.2) in which each water molecule is bound so tightly that it can neither move nor rotate freely. These intermolecular bonds are elastic (like springs), permitting molecules to vibrate but inhibiting long-range movements.

Because of its internal structure, ice is a rather open network of water molecules. The molecules in ice are not as closely packed as a similar number of molecules in liquid water, which fit together like marbles in a cup. Ice at 0°C has a density of about 0.92 g/cm^3. Liquid water at the same temperature has a density of about 1.0 g/cm^3. This density difference explains why ice floats in water. Most solids sink in their melts; this is another instance of water's unusual properties.

As we shall discuss later, sea ice contains less salt than does seawater. Despite the openness of the ice structure, most salt atoms cannot fit between the water molecules. Salts cannot substitute for water molecules in ice; most impurities (salts and gases) are excluded from ice when seawater freezes. (There will be more about this when we discuss sea ice later in this chapter.)

When ice melts, liquid water forms. After observing ice crystals disappear during melting, we might guess that all intermolecular bonds in the ice structure would also disappear. But this is not the case. Instead, many water molecules remain bonded together, forming **clusters** (Figure 3.3), which are surrounded by **unbonded water molecules.** Consequently, water is an unusual liquid; in fact, it behaves like a quasicrystalline substance at oceanic temperatures because of the tendency of water molecules to form hydrogen bonds. At low temperatures, many molecules in liquid water are in clusters at any given instant (Figure 3.3). The proportion of bonded molecules and the relative size of clusters decrease as water temperature rises (Figure 3.4). At water's boiling point, all bonds between molecules break down.

Liquid water's structure is dynamic. Its molecules alternate rapidly between structured and unstructured states. The intermolecular bonds that cause its clusters to hold together break and re-form many millions of times each second. In ice, these bonds persist much longer, breaking about once each second.

The relative abundances of structured and unstructured components of water change with varying temperatures, pressures, and salt contents. Rapid changes or "flickering" of the structure account for water's ability to flow. If its structure did not rapidly break and form again, water would be as rigid or brittle as ice, fracturing instead of flowing. Thus, liquid water, unlike ice, flows readily, maintaining a fixed volume at a given temperature.

Water vapor, a gas, has neither shape nor size and completely fills any container in which it is placed. For liquid water to change to water vapor, all inter-

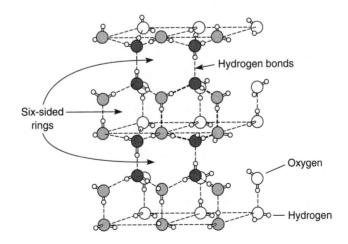

FIGURE 3.2
Crystal structure of ice. Note the six-sided rings formed by the water molecules and the hydrogen bonds linking water molecules in the ice structure.

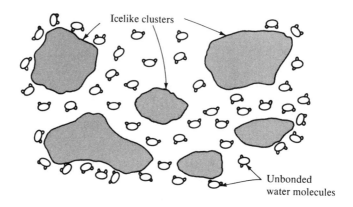

FIGURE 3.3

Schematic representation of the structure of liquid water. Note the structured icelike clusters surrounded by unbonded water molecules.

[After R. A. Horne, "The Physical Chemistry and Structure of Sea Water," *Water Resources Research* 1 (1965), p. 269]

FIGURE 3.4

Effect of temperature on the relative number of unbroken hydrogen bonds and the size of clusters in pure water.

[Data from G. Nemethy and H. A. Scheraga, "Structure of Water and Hydrophobic Bonding in Protein," *Journal of Chemical Physics* 36 (1962), p. 3394]

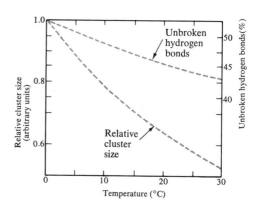

molecular bonds must be broken. Individual molecules may then move and rotate independently.

On a molecular level, water vapor in a container may be compared to a room full of bees. Each molecule, represented by a bee, moves independently, unaffected by the others. The pressure that a gas exerts on its container results from molecules colliding with the walls. As temperature rises, the molecules move more rapidly, and collide more frequently with the container walls. In other words, pressure increases with higher temperatures. In gases, molecules are widely spaced, and thus other gases can be added with little difficulty, although this increases total gas pressure.

3.3   THERMAL PROPERTIES

Changes in a substance's physical form, such as a solid changing to a liquid or to a vapor, are called **changes of state.** Changes of state require breaking of intermolecular bonds, or forming of new ones. If bonds are broken, energy is taken up; when new bonds are formed, energy is released, usually as heat.

A **calorie** (abbreviated cal) is defined as the amount of heat (or energy) required to raise the temperature of 1 g of liquid water by 1°C. For example, 100 cal must be supplied to 1 g of pure water to change its temperature from the melting point (0°C) to the boiling point (100°C).

To illustrate relationships between addition of heat and changes in temperature as well as changes of state in our water system, let us take 1 g of ice and heat it (Figure 3.5). Water molecules in ice can vibrate, because the molecular bonds are somewhat elastic. Energy supplied by warming the ice increases molecular vibrations and stretches the molecular bonds.

At its melting point (0°C), molecular vibrations are strong enough to break water molecules loose from the ice structure, forming liquid water. If no more heat is added, ice and liquid water will coexist in **equilibrium.** At equilibrium, the number of molecules gaining enough energy to break free of the ice structure at any instant is balanced by the number of water molecules losing energy and rejoining the ice structure. Unless energy is added or removed, the relative amounts of ice and liquid water remain unchanged.

Adding more heat melts more ice; removing heat (cooling) has the opposite effect. The last of the ice melts as we continue to supply heat (total about 80 cal/g). The amount of heat needed to melt 1 g of ice at 0°C is called the **latent heat of melting.** The heat consumed is released when water refreezes.

FIGURE 3.5
Temperature changes when heat is added to (or removed from) ice, liquid water, or water vapor. Note that the temperature of the system does not change in mixtures of ice and liquid water or liquid water and water vapor.

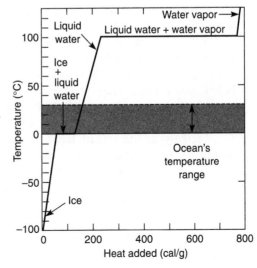

When the last bit of ice disappears, additional heating causes liquid-water temperatures to rise. Energy, no longer used for breaking bonds, causes water molecules to move more rapidly, and a rise in temperature can be measured. Between the melting and boiling points, addition of a fixed amount of heat causes a fixed temperature rise.

The relationship between the amount of heat supplied and the resulting temperature change is called **heat capacity.** (This same relationship was used to define the calorie.) About five times as much heat must be supplied to water to cause a 1°C rise in temperature than is necessary to cause a 1°C rise in temperature of an equal mass of rock. (You may notice this on a summer day at the beach. The ground heats up during the day but cools quickly at night, whereas water temperatures change little over a day.)

Liquid water's heat capacity is unusually high for a liquid because of its internal structure. Therefore, ocean waters can absorb or release large amounts of heat and yet undergo little temperature change.

When liquid water reaches 100°C (its boiling point at normal atmospheric pressure), many molecules have enough energy to break free, forming water vapor, a gas. A large amount of heat energy (539 cal/g) is needed to evaporate water at 100°C, and the reason for this is easily understood. When water evaporates, hydrogen-bonded clusters are broken up, but, because of the strong hydrogen bonds, this requires a great deal of energy. Hence, water's latent heat of evaporation is high.

In contrast, water's **latent heat of melting** at 0°C is only 80 cal. Because the structure is not completely destroyed, not all hydrogen bonds are broken (Figure 3.4). Therefore, melting of a given mass of ice requires about 1/7 as much energy as evaporating the same amount of liquid water.

Although water boils at 100°C, water vapor can also form from ice or liquid water at much lower temperatures, because some molecules gain enough energy to break their bonds and escape. You can verify this for yourself. A wet cloth can dry even when completely frozen, because ice sublimates; that is, ice changes to water vapor without going through a liquid state. Evaporation from the sea surface (extremely important to Earth's heat and water budgets) occurs well below water's boiling point. The average sea surface temperature is about 18°C (64°F).

Evaporation of water below the boiling point requires more heat per gram of water vapor than does evaporation of the same amount of water at its boiling point. This increase in the **latent heat of evaporation** is caused by the extra energy required to break hydrogen bonds at lower temperatures, as shown in Table 3.1.

These processes, or changes of state, are reversible. In other words, the latent heat of evaporation is recovered by condensing water vapor, forming liquid water. Condensing 1 g of water vapor at 20°C to form water at the same temperature releases 585 cal. This is why rain is sometimes called "liquid sunshine."

Evaporation effectively removes heat from the sea surface. This heat is returned, warming the atmosphere, when water vapor condenses to form rain or snow. Heat transport by water vapor accounts for the mild winters of humid

TABLE 3.1
Latent Heat of Evaporation of Water at Various Temperatures

Water Temperature (°C)	Latent Heat of Evaporation (cal/g)
0	595
20	585
100	539

coastal areas, such as the Pacific Northwest in the United States. Abundant rainfall releases heat in the atmosphere, preventing the much lower winter temperatures found in dry continental regions far from the ocean. (This is discussed further in Section 4.3.)

3.4   SEA SALTS

Seawater is a solution of salts, of nearly constant composition, dissolved in variable amounts of water. Water, its most abundant constituent, determines most of seawater's physical properties. (The amount of fresh water in seawater is increased by rain and other forms of precipitation and is diminished by evaporation.) The presence of salt in seawater influences its density and other physical characteristics. We consider some of these influences in this and subsequent sections.

Despite the large number of elements (more than 70) dissolved in seawater, only six (Figure 3.6) constitute more than 99% of all sea salts: chlorine (Cl) , sodium (Na), magnesium (Mg), calcium (Ca), potassium (K), and sulfur (S), which occurs in seawater as the sulfate ion (SO_4). Common table salt (sodium chloride) alone makes up nearly 86%. All occur as ions, electrically charged atoms, or groups of atoms called molecules.

When substances dissolve in water, electrons are stripped from some constituents and form positive ions; conversely, constituents gaining extra electrons are negatively charged. The numbers of positive and negative ions in a solution must balance. In seawater, the major constituents occur in two forms of ions. Positively charged ions, which lack electrons, include sodium (Na^+); and magnesium (Mg^{++}). Negatively charged ions, which have excess electrons, include chlorine (Cl^-); and sulfate (SO_4^{--}).

Oceanographers use **salinity** (grams of total dissolved salts present in 1 kg of water) to express the salt content of seawater. Salinity is commonly determined by measuring seawater's ability to conduct electricity. The higher the salinity, the less resistance (or greater conductivity) the seawater sample offers to electrical currents flowing through it. The four major ions in seawater, listed above, account for 98% of seawater's electrical conductivity.

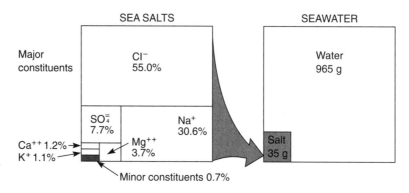

FIGURE 3.6
Relative proportions of water and dissolved salts in 1 kg of seawater, with a salinity of 35 parts per thousand or 3.5%.

3.5 SOURCES OF SEA SALTS

The major constituents in sea salts come primarily from three sources: volcanic eruptions; chemical reactions between seawaters and hot, newly formed, crustal rocks; and weathering of rocks on land (Figure 3.7). Sea-salt composition has remained nearly constant for hundreds of millions, probably billions, of years, controlled by various chemical and biological reactions. Thus, the rates of adding new salts must closely balance their removal rates from seawater. Otherwise, seawater would be changing with time, and we have no evidence of this.

Indeed, several lines of evidence indicate that seawater's composition has remained nearly constant for billions of years. Salt deposits formed when seawaters evaporate are similar in composition throughout the geologic record, indicating that there have been no major changes in sea salt composition. Also, shells formed by marine animals over the past 500 million years show no evidence of major shifts in the composition of sea salts. Thus, the question becomes, what controls the chemical composition of seawater within such narrow limits?

Chemical interactions between seawater and recently formed oceanic crust probably constitute the most important control over sea-salt composition. Magnesium and sulfate are removed by these water-rock interactions, whereas other materials and elements, such as silica, rubidium, and lithium, are added. It is estimated that all ocean waters circulate through newly formed ocean crust every 5 to 10 million years. Chemical reactions occurring there probably are the long-term controls on sea-salt composition.

Many constituents of sea salts come from weathering of rocks on land. As rocks decompose, forming soils, they release soluble constituents, such as phosphate and silica, which are carried to the ocean by rivers. River waters also contain calcium and bicarbonate (Figure 3.8) derived from limestone dissolution. Upon entering the ocean, dissolved salts remain behind while water itself con-

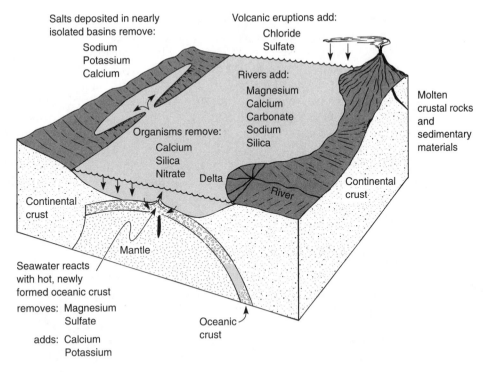

FIGURE 3.7
Schematic representation of the major processes controlling sea-salt composition.

FIGURE 3.8
Relative abundance of dissolved solids in average river waters.

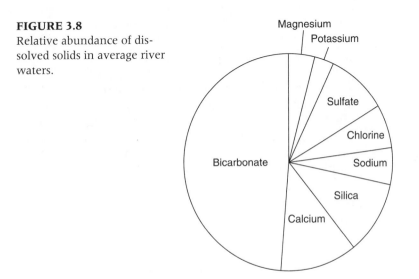

tinues to circulate through the atmosphere by evaporation and precipitation (rain and snow) and then returns back to the ocean.

Other processes also remove dissolved constituents from seawater. Some constituents are removed by chemical reactions with weathered mineral grains. Compounds needed for the growth of organisms, such as silica and phosphate, are removed by biological processes. (We discuss this further in Chapter 7.)

Volcanic eruptions release gases, which contribute both sulfate and chloride. These constituents are removed when seawater in isolated arms evaporates, depositing rock salt. These ancient salt deposits are the remains of ancient seas that dried up, often many times. Such salt deposits are especially likely to form during the early stages of ocean-basin formation, when the newly formed basins (such as the Red Sea) are still isolated from the rest of the ocean. The relative importance of these various removal processes cannot yet be assessed.

3.6   PARTICLES

Particles dispersed in seawater also affect chemical and biological processes (Figure 3.9). They can alter the chemical compositions of both seawater and sediments, which we will discuss further in Chapter 9. As they sink, particles can react chemically, removing reactive elements from near-surface waters and transporting them to deeper waters. Particles are also eaten by the many organisms that filter seawater to obtain their food. Particle surfaces also provide substrates where microscopic organisms such as bacteria can live. (Biological processes involving particles are discussed in Chapter 8.)

Most particles (30 to 70%) in seawater come from organisms. Skeletal remains (calcium carbonate and silica) from marine organisms alone make up 25 to 50% of seawater's particulate matter. (As we see later, these are major contributors to the sediment deposits on the ocean bottom.) Large, rapidly settling, biologically derived particles (sinking 100 m per day), reach the bottom in a few days. Many of these large particles are fecal pellets, voided by animals feeding in near-surface waters. Fecal pellets transport food produced in surface waters to organisms living in the deep ocean. These large particles settle out on the bottom, adding to sediment deposits.

Seawater also contains large numbers of much smaller particles that sink slowly, taking years to reach the ocean bottom. This allows ample time for chemical and biological changes to occur. Consequently, these slowly sinking particles influence the chemical compositions of the sea salts. Most of these tiny particles are fragments of plants and animals. Inorganic particles are brought to the ocean by rivers, by winds blowing over deserts and mountains, by meteoritic dust, and from large volcanic eruptions. These tiny particles provide spaces for microbial processes that influence the chemical composition. Thus, the large particles primarily affect the bottom organisms and sediment deposits. The tiny particles have their greatest influence on seawater composition.

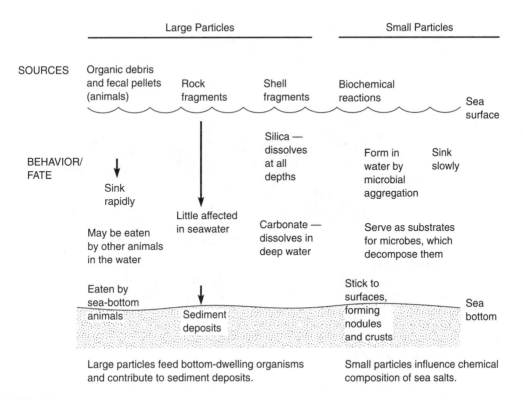

FIGURE 3.9
Schematic representation of particle behaviors in the ocean.

In the open ocean, soluble particles, often derived from plants and animals, dissolve before reaching the bottom, thereby changing the chemical compositions of deep-ocean waters. These particles transport nutrients (nitrogen, phosphate, and silicate compounds from organisms) that are necessary for plant growth from near-surface waters to subsurface waters. There the particles dissolve, releasing their nutrients to the surrounding waters. (The importance of this process, called the **biological pump,** to biological and chemical processes is discussed in Chapter 7.)

3.7   DISSOLVED GASES

Seawater also contains dissolved gases (Figure 3.10). Stirring of the sea surface by winds and waves facilitates dissolution of atmospheric gases in surface waters. Water of a given temperature and salinity is saturated with gas when the amount of gas dissolving in the water equals the amount leaving during the

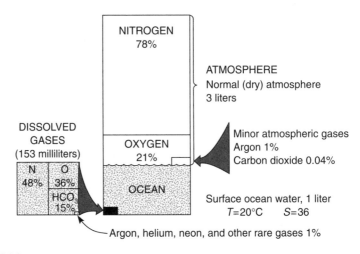

FIGURE 3.10
Gases dissolved in saturated seawater in equilibrium with a dry atmosphere. For each liter of seawater on Earth, there are 3 l of atmospheric gases.

same time. Surface seawater is normally saturated with atmospheric gases—oxygen, nitrogen, and trace gases.

The amount of gas that can dissolve in seawater is determined by its temperature and salinity. At higher temperatures or salinities, seawater can hold less gas; of the two, temperature is more important. Along with water temperature and salinity, the amount of gas dissolved in a parcel of seawater is controlled by the conditions where the water was last exposed to the atmosphere at the ocean surface.

Once water sinks beneath the surface, it can no longer exchange gases with the atmosphere. Assuming no biological processes are involved, the amount of gas in a parcel of water may remain unchanged, except by movements (diffusion) of gas molecules through the water (a very slow process) or by mixing with other water masses containing different amounts of dissolved gas. Generally, nitrogen and rare, chemically inert gases (argon, neon) in the atmosphere behave this way; we say that their concentrations are **conservative properties,** affected only by physical processes. (Salinity is another example of a conservative property of seawater.) Seawaters are normally saturated with nitrogen and rare gases, because these gases are unaffected by chemical or biological processes.

In addition to mixing and diffusion, some gases, primarily oxygen and carbon dioxide, are affected by biological or chemical processes that change their concentrations; the concentrations of these gases are examples of **nonconservative properties.** For instance, both oxygen and carbon dioxide are released and also used at varying rates in the ocean by photosynthesis and respiration, which we discuss in Chapter 8. Variations in the concentrations of these gases can be used to track movements of subsurface waters.

Compared with other atmospheric gases, carbon dioxide is present in unusually large amounts because of chemical and biological processes in the ocean that involve carbon dioxide. These chemical reactions control how acidic or alkaline seawater is. These same processes in the ocean also influence the amount of carbon dioxide that remains in the atmosphere as a result of burning of fossil fuels. Processes affecting carbon dioxide may also determine how Earth's climate responds to future carbon dioxide releases. (There will be more about this in Chapter 4.)

3.8   ALKALINITY AND ACIDITY

Seawater's acidity and alkalinity are important properties, especially for marine life. The amount of carbon dioxide dissolved in the water helps control the balance between the two. In the open ocean, this balance changes little, but it becomes important in coastal waters and in lakes—especially when they are polluted.

First we must define acidity and alkalinity. An **acid** is a hydrogen-containing compound that releases its hydrogen ions (H+) when dissolved in water. Strong acids readily release their hydrogen ions; weak acids do not release them so readily. A **base,** on the other hand, is a compound that releases hydroxyl ions (OH⁻). Like acids, some bases are weak and others are strong.

Water acts as both an acid and a base as water molecules break up and reform as shown below:

$$H_2O \quad <<<>>> \quad H^+ \quad + \quad OH^-$$
water   forms   hydrogen ions   plus   hydroxyl ions

Pure water is neutral; in other words, there are as many acid ions as base ions. If we add acid, the solution becomes acidic. Conversely, the solution becomes basic if we add a base.

To indicate water's relative acidic or basic nature, we use the **pH scale,** which is based on the abundance of hydrogen ions, as shown below:

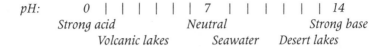

pH: 0 | | | | | | 7 | | | | | | 14
        Strong acid   *Neutral*   *Strong base*
            Volcanic lakes   *Seawater*   *Desert lakes*

We assign a pH value of 7 to neutral water. The most acidic solution would have a pH of 0 and the most basic would have a pH of 14. Seawater normally has a pH of about 8.1 (slightly basic) because of the carbon dioxide dissolved in it. Carbon dioxide interacts with calcium carbonate in seawater to maintain its narrow range of pH values. This process is called **buffering.** Lake waters exhibit larger ranges of pH because they are not as well buffered as seawater. Acidic lakes lack enough calcium carbonate to balance the acids contained in acid rains. Desert lakes are often basic because they contain large amounts of dissolved carbonate, brought in by streams.

3.9   PHYSICAL PROPERTIES OF SEAWATER

Some properties of seawater change as salt concentration increases. For instance, changing salinity from 0 to 40 causes **viscosity** (internal resistance to flow) to increase about 5%. Adding sea salts to water also changes its temperatures of maximum density and initial freezing. Because salt does not fit into the ice crystal structure, it inhibits ice formation and depresses the initial freezing point (Figure 3.11). Adding salt causes the mixture to freeze at temperatures below 0°C.

Seawater does not freeze completely at a fixed temperature as does pure water; in other words, seawater has no fixed freezing point. As seawater freezes, salts are excluded from the ice structure. Consequently, the unfrozen water becomes saltier and therefore freezes at still lower temperatures. Unless cooled to very low temperatures, some concentrated brine remains.

Processes that depress water's initial freezing point also depress its temperature of maximum density. Sea salts inhibit development of the clusters that cause pure liquid water to expand near the freezing point. Adding sea salt to water lowers the temperature of maximum density. At a salinity of 24.7, the maximum density and the initial freezing temperature occur at –1.33°C. At salinities greater than 24.7, seawater does not exhibit a density maximum (Figure 3.11). Hence, typical seawater (salinity of 35) becomes progressively denser as it cools, until it begins to freeze.

3.10   SEAWATER DENSITY

Temperature, salinity, and pressure control seawater **density.** Of the three, temperature and salinity are the most important. In the open ocean, seawater density varies only a small amount. As we shall see in Chapter 5, these slight density differences cause ocean currents. Consequently, oceanographers must determine seawater density with great precision. Normally, seawater density is calculated from precise measurements of the temperatures (accurate to within 0.002°C)

FIGURE 3.11
Effects of salinity on the temperatures of maximum density and initial freezing of seawater. At salinities below 24.7, water goes through a density maximum before it freezes. At higher salinities, seawater simply begins to freeze without exhibiting a density maximum.

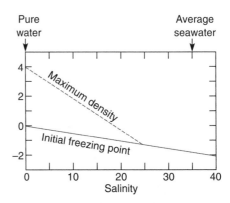

FIGURE 3.12

Changes in seawater density (in grams per cubic centimeter) are caused by variations in salinity and temperature. [After U.S. Navy, *Instruction Manual for Oceanographic Observations*, H. O. Pub. 607 (Washington, DC, 1955), p. 42]

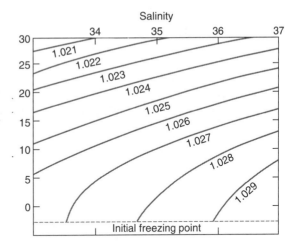

and salinities (accurate to within 0.002) of water samples. From these measurements, density is calculated to an accuracy of one part in 500,000.

Figure 3.12 shows maximum water-density changes for salinities between 30 and 37 and temperatures from −3 to 30°C. This encompasses the temperature and salinity ranges of the entire ocean. Figure 3.12 also shows the relative effects of temperature and salinity on seawater density. At 30°C, a change in salinity from 34 to 35 changes the density from 1.021 to about 1.022. Density is changed an equal amount by cooling water with a salinity of 37 from 27.5 to 24.3°C, a change of 3.2°C. Such temperature changes occur commonly at the ocean surface. Large salinity changes usually occur near land, as a result of river discharges and increased precipitation there, and in polar regions, because of ice formation.

3.11   SEAWATER STABILITY

The relative density of a particular bit of seawater controls the depth at which that water parcel occurs in the ocean. Changes in density, resulting from processes occurring at the ocean surface, cause the sluggish, deep-ocean currents.

To demonstrate the effects of different water densities, conduct the following experiments. First, fill a glass with tap water. Fill a medicine dropper with salty water colored with ink or food coloring. Put a drop of colored salty water into the glass of fresh water. The salty water sinks to the bottom because it is denser than the fresh water. Conversely, a drop of colored fresh water put into a beaker of salty water remains at the surface because it is less dense.

For the next experiment, create a two-layered system (Figure 3.13). Make dense, salty water by dissolving as much salt as the water in a half-filled container will hold. (Add salt until some remains undissolved even after vigorous stirring.) Then carefully pour tap water on top of the salty water. Wait until the water

FIGURE 3.13
Vertical movements of inter-
mediate-density waters in a
simple, stable, two-layered
system. If the water drop is
denser than its surroundings,
it sinks; if it is less dense, it
rises.

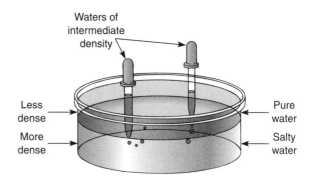

movements caused by adding the tap water have subsided. Now you have a stable, two-layered system where the denser salty water lies below the fresh water.

The system is stable and will remain unchanged unless you mix it by stirring
or heating. Stability is resistance to change in the system. If you slightly disturb a
stable system, it will return to its initial state after the disturbance ceases.

If you now add slightly salty, colored water to our two-layered system, it
comes to rest at an intermediate level. The exact level will depend on the
changes of density with depth and the density of the colored salty water. There
is a stable density distribution if the most dense water (in this case the saltiest) is
at the bottom and the least dense water is on top. Waters of intermediate density
will be at intermediate depths.

When a drop of salty water is added, it creates an **unstable density distribution.** In this case, the drop is denser than surrounding waters and it sinks.
Conversely, if a dropper filled with slightly salty water is carefully placed down
in the very salty water and a drop is released, it rises because you again have
created an unstable density distribution; the water drop is less dense than its surrounding waters. Thus, an unstable system spontaneously moves toward a stable
density distribution.

You can make a **neutrally stable density distribution** by mixing the
water so that density is the same throughout. In fact, if the original two-layered
system stands long enough, salt will diffuse throughout the water, eventually
equalizing water density in the container. In this case the system does not return
to its initial state after a disturbance. Neutrally stable systems are easily mixed.

Effects of temperature variations on water density can be shown by heating
one side of a dish with a flame (Figure 3.14). The warmed water becomes less
dense and rises. Cooler waters around the ice cube sink on the other side of the
container and flow along the bottom, replacing the rising waters.

These density-controlled vertical movements are called **convection currents.** Convection is common in the atmosphere, which is warmed at the bottom and cooled at the top. Ocean waters, however, are heated and cooled at the
top, except over recently erupted ocean-floor lava. Consequently, convection is
relatively rare in the ocean, occurring in only a few locations (primarily polar

FIGURE 3.14
Convection currents caused by warming water at the bottom and cooling it at the top.

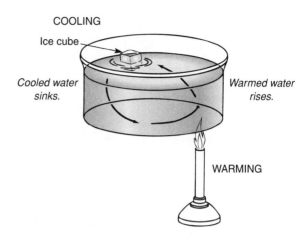

COOLING

Ice cube

Cooled water sinks.

Warmed water rises.

WARMING

oceans, and around recently emplaced ocean-floor lava flows), and even there only intermittently.

The ocean can be simulated experimentally by shining a heat lamp on the surface of a shallow dish of cool water. This creates a layer of warm, less dense water on top, which is a stable density distribution. Vigorous stirring is required to mix such a system. (We see the effect of this in Chapter 4.)

3.12   SEA ICE

Sea ice forms in coastal and high-latitude oceans in winter. When seawater is chilled below its initial freezing point, microscopic ice crystals form and later grow into hexagonal needles 1 to 2 cm (about a half an inch) long. At this stage, the sea surface becomes dull and no longer reflects the sky. As freezing continues, ice crystals freeze together, covering the ocean surface like a blanket of wet snow. Eventually these crystals grow downward and freeze together. A thin, plastic layer of ice then forms, which includes small enclosed cells containing seawater. These ice films are easily moved by winds.

Ice crystals themselves contain no salt, but brines in small cells between them may be saltier than seawater. Typically 1 kg of newly formed sea ice consists of about 800 g of ice (salinity 0) and 200 g of seawater (salinity 35); the average salinity of newly formed sea ice is about 7.

As the temperature continues to fall, more ice forms under the ice layer, making it thicker. Brines in the cells also partially freeze, adding ice to interior cell walls. The remaining brines become saltier. If the temperature goes low enough, salt eventually crystallizes as the last bit of water freezes.

The salt content of newly formed sea ice depends on the temperature. At temperatures near freezing, ice forms slowly, allowing brines to escape. Little seawater remains in the cells, and the ice therefore contains little salt. At lower temperatures,

FIGURE 3.15
A U.S. Coast Guard icebreaker moves through first-year ice near Antarctica. (Courtesy NASA)

ice forms more rapidly, trapping seawater. In this case, the sea ice contains more salt, but it is always less salty than the seawaters from which it formed.

As sea ice ages, it excludes salt and hardens. An ice layer up to 20 cm (about 8 in.) thick can form in one winter; this is called first-year ice (Figure 3.15). First-year ice dominates the Southern Ocean around Antarctica, where there is little multiyear ice. But in the Arctic pack ice of the central basin, sea ice melts little during summer. There multiyear ice dominates.

Over many seasons, the maximum ice thickness (multiyear ice) is usually 2 to 3.5 m (about 7 to 11 ft). Winds can pile pieces of ice on top of one another, forming pressure ridges. Such pressure ridges can extend many meters below the ice pack and constitute hazards to submarines navigating under the Arctic ice or to icebreakers traveling through the ice pack.

3.13   SOUND IN THE OCEAN

Seawater is nearly transparent to sound. Thus, sound is used by marine animals and (more recently) by humans to locate targets, to communicate through the ocean, and to explore oceanic depths. Two examples of this are the echo sounder, which revolutionized mapping of the ocean bottom when introduced on research vessels in the 1920s, and fish locators, which greatly improved fishermen's ability to locate schools of fishes.

Average sound velocity in the ocean is 1450 m/s (0.9 mi/s). Higher water temperatures, salinities, and pressures all increase sound velocity. At a depth of around 1 km, the combination of these three factors, primarily controlled by temperature variations, causes a minimum in sound speed. Thus, sounds originating in or below this layer, called the **sound channel,** will be deflected into this channel. Because of this feature, sounds tend to remain in the sound channel. This, combined with seawater's transparency to sound, makes possible long-distance communications in this layer (Figure 3.16). The sound channel was once proposed as a way for aviators downed at sea to signal their location by detonating a small explosive charge in the sound channel, which would be heard by listening stations ashore. One could also determine the signal's location by comparing arrival times at the various stations.

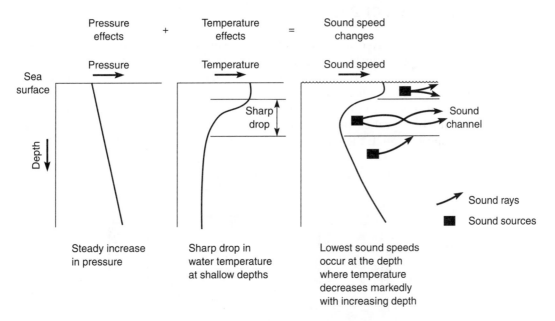

FIGURE 3.16

Sound speed in the ocean is controlled by pressure, temperature, and density. Salinity changes little over most of the ocean and is thus ignored. Pressure increases steadily with depth and so does sound speed. Temperature decreases markedly relatively close to the ocean surface and then decreases slowly with greater depth. Temperature effects dominate in the upper part of the water column and pressure dominates in the deeper waters. As a result, a sound speed minimum (called the **sound channel**) occurs in the depth zone where water temperature decreases most rapidly. Sounds are trapped in the sound channel and thus can be detected at great distances.

Sound pulses are now used to determine average ocean water temperatures over entire basins. Repeated observations using different combinations of source and receiver arrays can also be used to detect subsurface water masses and to track their movements.

Plans have been made to use such sound arrays for monitoring of the ocean depths to detect possible changes in water temperatures and thus determine if ocean waters are indeed warming. One proposal is to install such transmitters and receivers so that the temperature of the deep ocean can be systematically measured to determine whether Earth's temperature is indeed rising, as predicted by many scientists on the basis of changes in earth's atmosphere.

During the Cold War, both the U.S. and Soviet navies used elaborate listening arrays to detect and track enemy submarines. After the Cold War ended, the U.S. Navy permitted civilian scientists to use these sensitive listening devices. Thus, scientists can determine times and locations of sea-floor volcanic eruptions, as we discussed in Chapter 2.

3.14   WATER CYCLE

Water constantly evaporates from the ocean surface, leaving behind sea salts. Most of this water (about 91%) falls as rain on the ocean surface (Figure 3.17). The remaining 9% is carried by winds and falls as rain or snow on the land and of this precipitation about 70% evaporates. Only about 30% of the precipitation on land returns to the sea through river runoff. This is called the **water cycle;** it is sometimes called the hydrologic cycle. As we have already discussed, water transport through the atmosphere is also part of Earth's heat budget.

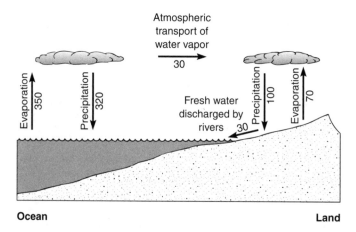

FIGURE 3.17
Schematic representation of Earth's water cycle, showing movements of water (in thousands of cubic kilometers per year) from the ocean surface to the continents by winds and then back to the ocean by rivers.

Some water falls as snow, forming ice on glaciers, where it may remain for thousands of years as the glacial ice moves slowly back to the ocean. A small amount of precipitation flows into permeable rocks and becomes ground water. It takes decades to centuries for this water to flow into rivers or the ocean. Only a tiny fraction of Earth's water is in the atmosphere or in fresh-water reservoirs at any instant.

3.15   RESOURCES FROM SEAWATER

Seawater itself supplies us with several materials, such as fresh water, salt, bromine, and magnesium. It is an attractive source of raw materials for several reasons. First, because of the abundance of seawater, extraction of raw materials is not likely to deplete the ocean. Furthermore, several substances are added to the ocean in amounts that equal or exceed the amounts extracted.

Water is, of course, the ocean's most abundant resource. Some coastal cities in arid regions derive fresh water from the ocean by evaporating seawater. The problem is to recover the water where and when it is needed and at affordable costs. Desalinization of seawater is common in large cities around the Arabian Gulf, where fresh water is extremely scarce but oil and gas are plentiful.

Despite the wide variety and abundance of other materials dissolved in the ocean, the extremely dilute nature of seawater makes it expensive to produce them. Gold is one example. It occurs in seawater at levels of about 4 g of gold per million tons of seawater. This amounts to about 5 million tons of gold in the ocean. The cost of pumping seawater and extracting the gold, unfortunately, would greatly exceed its value.

3.16   SALT TRADE

Salt is the other substance that comes primarily from the ocean, either directly or indirectly. Producing and trading of salt are among the oldest known human commercial activities. Indeed, many believe that the earliest known roads were built for transporting salt. Many ancient cities arose as centers of the salt trade. Wars and revolutions have been caused by needs for salt or by attempts to monopolize salt sources. Humans require salt in their diet, to cure meats (before refrigeration), and as a raw material for the chemical industry and even such mundane tasks as melting ice from sidewalks and streets.

Seawater was long the dominant source of such salt, although ancient rock salt deposits are often now exploited, in part because they are often more conveniently located and/or provide purer supplies. Salt is usually recovered by evaporating seawater in shallow enclosed bodies of water—called salt pans—along sunny, low-lying coasts. Salt pans (Figure 3.18) operate at the South end of San Francisco Bay (identified by their bright red color when viewed from above) and on the dry south coast of Puerto Rico. Salt production is a common activity

FIGURE 3.18
Shallow basins (salt pans) are used to evaporate seawater for production of salt in south-
ern San Francisco Bay.
(Courtesy Leslie Salt Company)

along many coastlines. Even some northern coastlands, such as Newfoundland
and Nova Scotia in Canada, produced sea salt for centuries to preserve fish for
transport back to Europe.

Northern Germany and China had salt deposits that were exploited. Some
were mined to produce rock salt; in others, salt springs draining the salt deposits
were collected and evaporated to recover the salt. The Chinese developed
drilling techniques that permitted them to recover deeply buried brines, which
were then evaporated.

QUESTIONS

1. Using water as an example, describe the three forms of matter.
2. Describe the atomic structure of water. Why is water called a polar molecule?
3. How much heat (in calories) must be removed per square centimeter of water surface to cool by 1°C a water column 100 m thick?
4. Explain why ice is less dense than liquid water.
5. Define latent heat. Why is the latent heat of evaporation of water greater than the latent heat of melting?
6. Explain why fresh water is densest at 40°C.
7. How much heat (in calories) is required to convert 5 g of ice at 0°C to liquid water at 5°C?
8. Why does seawater not freeze completely at a fixed temperature?
9. List, in order of abundance, the six most common constituents in seawater.
10. What three factors control seawater density?
11. Explain why carbon dioxide makes up 15% of the gases dissolved in seawater but only 0.04% of the gases in the atmosphere.
12. Why are the dissolved-oxygen concentrations in seawater considered to be nonconservative properties?
13. Define salinity.
14. How does circulation of seawater through newly formed oceanic crust alter the composition of seawater?
15. Describe how sea ice forms.
16. Why is salt such a necessary commodity for humans?

SUPPLEMENTARY READINGS

Books

Davis, Kenneth S., and Day, John Arthur. *Water: The Mirror of Science.* Garden City: Anchor Books, Doubleday and Company, Inc., 1961. Elementary, nontechnical.

Deming, H. G. *Water: The Foundation of Opportunity.* New York: Oxford University Press, 1975. Elementary.

Articles

Gabianelli, J. J. "Water—the Fluid of Life." *Sea Frontiers* 62(5):258–270.

MacIntyre, Ferran. "Why the Sea Is Salt." *Scientific American* 223(5):104–115.

Revelle, R. "Water." *Scientific American* 209(3):93–108.

KEY TERMS AND CONCEPTS

Polar molecule
Hydrogen bonds
Forms of matter: solid, liquid, gas
Water-molecule clusters
Unbonded water molecules
Changes of state
Calorie

Equilibrium
Latent heat of melting
Heat capacity
Latent heat of evaporation
Sea salts
Salinity
Particles in seawater

Biological pump
Dissolved gases
Conservative properties
Nonconservative properties
Alkalinity and acidity
Acid
Base
Buffering
Viscosity
Temperature of initial freezing
Freezing-point depression
Temperature of maximum density

Density of seawater
Stability of seawater
Density distributions: stable, unstable, neutrally
    stable
Convection
Sea ice
Pressure ridges
Sound in seawater
Factors affecting sound velocity
Sound channel
Water cycle
Salt trade

4
Open Ocean and Climate

Far from land, incoming solar energy and winds dominate open-ocean processes. Here we consider first the effects of processes discussed in previous chapters. In Chapter 8, we will consider coastal oceans, where many oceanic processes discussed here interact with river discharges, complex shorelines, rugged bottom topographies, and tides and winds to form yet more complicated but short-lived currents.

4.1   LIGHT AND HEAT

Incoming energy from the Sun supplies heat and light to the ocean's surface. Some of the incoming solar radiation, called **insolation**, penetrates as light into the surface ocean layers. There it supplies the energy needed for plants to grow and provide food for marine animals. (We will learn more about this in Chapter 7.) The warm ocean surface also warms the overlying atmosphere, powering the atmospheric circulation.

Most of the incoming infrared energy from the Sun is absorbed within a few centimeters of the surface and changed into heat. About 60% of the visible light is absorbed within the first meter of seawater, and about 80% in the first 10 m (33 ft). Little light penetrates below about 140 m (460 ft), even in the clearest waters. Blue-green colors penetrate deepest in clear waters. In highly polluted waters or where marine plants are growing in profusion, all the light may be absorbed within a few centimeters of the water surface. Here, yellow light penetrates deepest into the water.

4.2   LAYERED STRUCTURE

Seawater temperature and salinity distributions control the behavior of open-ocean waters. Their distribution results from absorption of incoming solar radiation in near-surface waters and from heat and water-vapor transport over Earth's surface. Some incoming solar energy heats the ocean surface, but much of it goes into evaporation of water, supplying energy to power the atmosphere. The ocean and atmosphere together form an inefficient, Sun-powered engine that changes most of the incoming solar energy into winds and ocean currents.

The open ocean consists of three layers: the surface, pycnocline, and deep zones (Figure 4.1). Characteristics of the surface zone change because of seasonal variations in heating, cooling, evaporation, and precipitation. In polar and subpolar waters, freezing surface seawater forms sea ice in winter, most of which melts during the following summer.

The **surface zone** contains the ocean's least dense water, mostly because of its higher temperatures. (Salinity changes little in open-ocean waters.) The thickness of the surface zone is controlled by mixing, primarily wind mixing. Extensive mixing of water within the surface zone results in its nearly neutral stability, and thus water particles can easily move vertically. This zone is also frequently called the **mixed layer.** Surface waters have ample opportunity to adjust to local conditions, so their temperatures change seasonally along with local conditions.

Easy transfer of heat near the surface results in a nearly isothermal (iso = equal; thermal = heat) mixed zone. **Thermoclines** (thermo = heat; cline = slope), where temperatures change markedly with depth, separate Sun-warmed surface layers from colder waters below.

The **pycnocline zone** (pycno = density; cline = slope) also lies below the surface zone. There, water density changes markedly with depth (Figure 4.2). Because of these large density changes, pycnocline waters are very stable (they

FIGURE 4.1
The open ocean has a three-layered structure (shown schematically). The least dense waters are near the surface, and the densest waters are near the bottom. The relative volume of each zone is shown. Note that the waters of the deep zone are at the surface only in polar regions. Also note that the pycnocline rises near the equator.

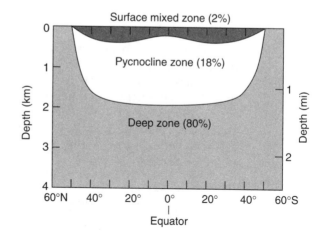

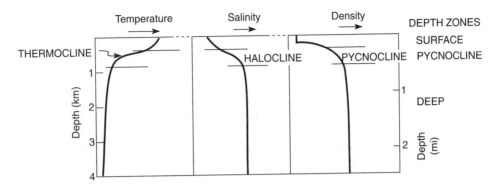

FIGURE 4.2
Vertical distributions of temperature, salinity, and water density in the open North Pacific.
[After J. P. Tully, "Oceanographic Regions and Assessment of Temperature Structure in the Seasonal
Zone of the North Pacific Ocean," *Journal of Fisheries Research Board* 21 (Canada: 1964), p. 942]

cannot easily move vertically but can move horizontally). The pycnocline thus
acts as a barrier to vertical water movements and serves as a floor to the surface
circulation with its seasonal changes in temperature and salinity. (In most parts
of the ocean, thermocline and pycnocline coincide; in other words, temperature
controls density there.)

In addition, the pycnocline is the ceiling for the **deep zone,** preventing
deep-ocean waters from readily mixing vertically with surface waters or equili-
brating with the atmosphere. Only in high latitudes and in polar areas (where
there is usually no pycnocline) are deep waters exposed to the atmosphere and
able to exchange gases (Figure 4.1). Waters below the base of the pycnocline
have an average temperature of only 3.5°C (38°F).

4.3   WATER MASSES

Large volumes of seawater move through the oceans as discrete **water masses,**
identifiable by their characteristic temperatures and salinities. These water
masses form at the ocean surface, and their temperatures and salinities reflect
surface conditions where they formed. If a newly formed water mass is denser
than its surrounding waters, it sinks to a level determined by its density relative
to the density distribution in the nearby ocean. Below the surface, water masses
are moved by subsurface currents, often for thousands of kilometers. After hundreds of years (possibly 1000 years), the deep waters return to the ocean surface,
again to exchange gases with the atmosphere and to be warmed by heat from
the Sun. Oceanographers trace subsurface water-mass movements, using
changes in dissolved gas concentrations, especially dissolved oxygen, and the
presence of pollutants from nuclear weapons testing and even atmospheric pollutants, such as chlorinated hydrocarbons.

The densest water masses in the ocean form in polar regions, where waters of moderately high salinity are intensely cooled at the ocean surface. (This accounts for the low water temperatures below the pycnocline.) These processes increase the depth of the pycnocline by the sinking of dense waters from the surface. If dense enough, these water masses may sink all the way to the bottom and flow along the ocean floor. A water mass of intermediate density will flow at a level appropriate to its density between the denser bottom waters and the less dense waters of the surface zone. The vertical position of a water mass of intermediate density is like the position of a card in a deck of cards; the layers below have greater densities, and the ones above have lower densities.

4.4 TEMPERATURE: HEATING AND COOLING

Heating of the ocean surface occurs during daylight; it is generally warmest in late afternoons. The amount of energy the ocean absorbs depends on local cloud cover and the Sun's altitude. The Sun's altitude depends in turn on latitude and the time of year (Figure 4.3). More energy is absorbed when the Sun is high in the sky and less is absorbed when it is near the horizon. In the tropics and subtropics, the Sun is well above the horizon at noon in all seasons. Near the poles the Sun is never far above the horizon, and thus polar and subpolar regions receive much less insolation. Consequently, Earth's surface is heated by the Sun in the tropics and subtropics and is cooled by radiating energy back into space, primarily from the polar and subpolar regions (Figure 4.4).

Energy from the Sun at the top of the atmosphere averages about 0.5 cal/cm^2 of Earth's surface per minute. After passing through the atmosphere, the incoming solar radiation is 0.25 cal/cm^2 per minute, averaged over 24 hours at Earth's surface.

If the average insolation were absorbed and remained in the upper 1 m (3 ft) of the surface zone, water temperatures would increase about 3.5°C (6.3°F)

FIGURE 4.3
Variations in incoming solar radiation per unit area result from the different angles at which radiation strikes Earth's surface. Conditions are shown for northern summer, when the Northern Hemisphere receives more solar radiation than the Southern Hemisphere. The conditions are reversed in southern summer.

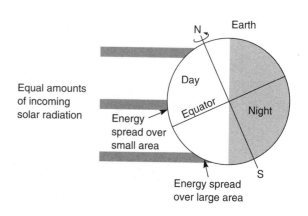

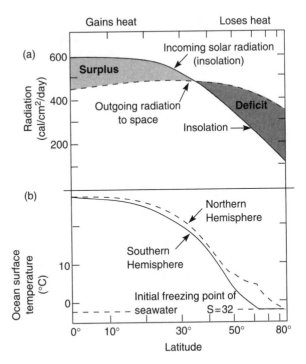

FIGURE 4.4
(a.) Radiation balances in the Northern Hemisphere, and (b) average surface ocean water temperatures in both hemispheres.
[Heat budget after H. G. Houghton, "Annual Heat Balance, Northern Hemisphere," *Journal of Meteorology* 11 (1954), p. 7; ocean temperatures from W. E. Forsythe (ed.), *Smithsonian Physical Tables,* 9th ed. (Washington, DC: Smithsonian Institution, 1964), p. 726]

in one day. However, the observed average daily temperature variation in the open ocean is only 0.2 to 0.3°C (0.4 to 0.5°F). This means that each day's input of solar energy is quickly mixed through near-surface waters or lost through evaporation. Because heat gained by the ocean during the day is distributed through a fairly thick surface zone, it is not readily lost at night. This mixing, combined with water's relatively high heat capacity, prevents large daily changes in surface-water temperatures. On land, heat remains near the surface (rocks do not conduct heat as well as water) during the day and is therefore readily lost at night. For these reasons, daily temperature ranges on land are much greater than those over the ocean.

If the ocean retained all the heat it absorbed, ocean waters would reach the boiling point in less than 300 years; obviously, this has not happened. Furthermore, fossil remains of ancient marine organisms show that ocean surface temperatures have changed little in the last billion years. This means that the ocean loses as much energy as it absorbs from insolation on an annual basis, as shown in Table 4.1. There may be long-term storage or release of heat from the ocean, but this is an unanswered question.

Heat is lost from the ocean surface day and night and in all seasons. Three processes are involved: (1) radiation of heat into space; (2) heating of the atmosphere through conduction; and (3) evaporation of water. About 40% of the insola-

TABLE 4.1
Heat Budget of the Ocean Surface (24-hour average)

Process	Heating	Cooling
	(cal/cm²/min)	
Incoming solar radiation (insolation)	0.25	
Radiation back to space		0.10
Evaporation		0.13
Atmospheric warming		0.02
Totals	0.25	0.25

tion received by the oceans radiates back to space, as shown in Table 4.1. Because Earth has relatively cool surface temperatures, it mostly radiates infrared (heat) back to space instead of visible light, which the Sun's much hotter surface radiates.

Some of the ocean's heat loss goes directly into warming the atmosphere by conduction, just as a pan is heated on a hot stove. Usually, ocean surfaces are about 1°C (1.8°F) warmer than the overlying air. About half the ocean's heat is lost to evaporating water, and later returns through the atmosphere as rain or snow.

Unequal heating of Earth's surface causes large differences in surface-water temperatures between tropical and polar regions (Figure 4.4). The ocean is warmest (25 to 30°C, or 77 to 86°F) in the tropical and subtropical regions and coldest (as low as −1.7°C, or 29°F) near the poles (Figure 4.5). As you can see, belts of equal surface-water temperature generally run east-west.

Surface isotherms (lines connecting areas of equal temperatures) deviate from their general east-west trends near continents. This is especially obvious in the western North Atlantic and North Pacific oceans. These deviations are caused primarily by ocean boundary currents, which generally parallel the shorelines. Some boundary currents, such as the Gulf Stream, transport warm water toward the poles. Others, such as Canada's Labrador Current, transport cool water toward the equator.

Ocean-surface temperatures (Figure 4.4) in the Northern and Southern hemispheres show similar changes in temperature with latitude in both hemispheres. The annual differences in surface temperature are greatest in the subpolar oceans at latitudes around 60°N and 60°S, and in the land-dominated Northern Hemisphere, and are least in the ocean-dominated Southern Hemisphere.

The lowest water temperature (−1.7°C, or 29°F) in the open ocean coincides with the initial freezing temperature (Figure 3.12) of seawater at salinity 32. Freezing or melting of sea ice in polar oceans acts as a thermostat, essentially fixing the lower temperature limit for surface water. Local surface-water temperature cannot go higher until sea ice melts, and as long as some surface water remains unfrozen at the same salinity, water temperature cannot go lower.

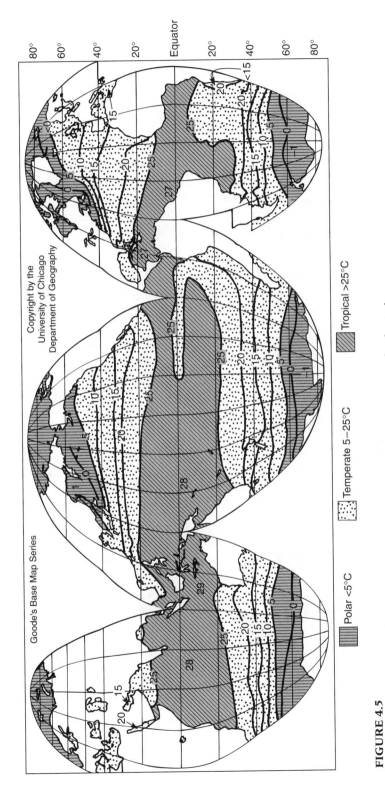

FIGURE 4.5

Ocean surface temperatures (°C) in northern winter. Note the east-west trends of equal-temperature zones.

[After H. J. McLellan, *Elements of Physical Oceanography* (Oxford: Pergamon Press, Ltd., 1965), p. 44]

Polar <5°C   Temperate 5–25°C   Tropical >25°C

Goode's Base Map Series

Copyright by the
University of Chicago
Department of Geography

4.5 SALINITY: EVAPORATION AND PRECIPITATION

Open-ocean salinity varies much less than does temperature (Figure 4.6). Salinity changes are caused primarily by evaporation (removal of fresh water as water vapor), by precipitation as rain or snow, and by river discharges (all adding fresh water).

In high latitudes, sea-ice formation plays another important role in controlling water density, because nearly fresh water is frozen into the ice, leaving behind most of the dissolved salts (see Chapter 3). These processes act on the ocean surface along with heating and cooling processes (Figure 4.7). Marked changes in salinity with depth (Figure 4.2) form the **halocline** (halo = salt).

Salinity changes affect seawater density. A change in salinity of 1 causes a greater density change than does a temperature change of 1°C (1.8°F), which means that in those parts of the ocean where surface waters (Figure 4.7) are diluted by excess precipitation (Figure 4.8), the main pycnocline and halocline often coincide. Nevertheless, over most of the ocean, despite important local effects of reduced surface salinity, the pycnocline is controlled by the thermocline. This is primarily a result of the relatively large temperature range (−1.7 to 30°C, or 29 to 86°F) of surface seawaters (Figure 4.6). In contrast, the salinity range for most of the ocean is small (33 to 37).

The amount of water evaporated from the ocean surface each year is equivalent to a layer about 1 m (3.3 ft) thick. About 90% of this water returns to the ocean surface as rain. The remainder falls as rain (or snow) on the land. Eventually this last 10% also returns, by means of rivers, to coastal oceans, where it causes lower salinities (Figure 4.7).

Evaporation from the ocean surface is controlled by (1) local insolation, (2) wind speeds, and (3) the relative humidity of the overlying air. Because of abun-

FIGURE 4.6

Water temperature and salinity ranges in the open ocean. [After R. B. Montgomery, "Water Characteristics of Atlantic Ocean and of World Ocean," *Deep Sea Research* 5 (1958), p. 144]

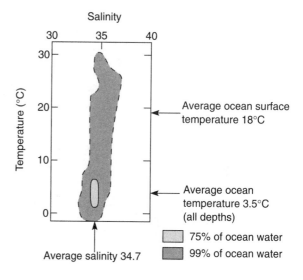

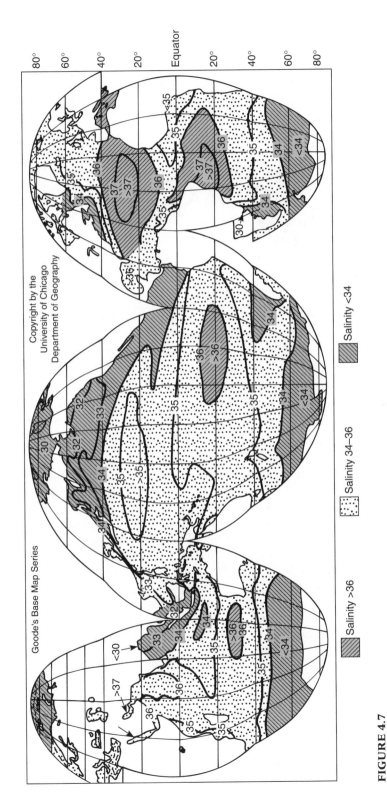

FIGURE 4.7

Ocean-surface salinities (expressed in parts per thousand). Lowest salinities occur near coasts, and highest salinities are found in the centers, or near their western margins, of gyres.

[After H. U. Sverdrup, M. W. Johnson, and R. H. Fleming, *The Oceans: Their Physics, Chemistry, and General Biology* (Englewood Cliffs, NJ: Prentice-Hall, Inc., 1942)]

Goode's Base Map Series

Copyright by the
University of Chicago
Department of Geography

Salinity >36   Salinity 34–36   Salinity <34

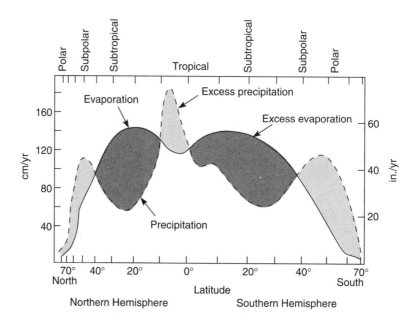

FIGURE 4.8
Distribution of evaporation and precipitation over the ocean.
[Data from G. Wüst, W. Brogmus, and E. Noodt. "Die Zonale Verteilung von Salzgehalt, Nieder-
schlag, Verdunstrung, Temperatur und Dichte an der Oberfläche der Ozeane," *Kieler Meeresforschun-
gen* V (1954), p.146]

dant insolation in the tropics and subtropics (Figure 4.3), large amounts of water
are evaporated there (Figure 4.8). Conversely, in polar regions, evaporation is much
less, and the excess precipitation dilutes surface waters, lowering their salinities.

Maximum evaporation occurs in subtropical regions (around 30°N and
30°S) under the nearly constant Trade Winds. The subtropics are also areas of
clear skies (high insolation) and relatively dry air. The relatively high surface
salinities (greater than 35) near 30°N and 30°S demonstrate that evaporation
dominates in these areas (Figure 4.8).

Diminished evaporation in equatorial regions is due in part to the light and
variable winds that give the region its name, the **Doldrums.** Extensive cloudi-
ness also diminishes insolation. Precipitation is also abundant near the equator,
as well as in the high latitudes, and in coastal oceans (Figure 4.8); in all these
areas, surface-water salinities (Figure 4.7) are markedly less than average ocean
salinity (34.7).

4.6   OCEAN CLIMATE

Oceanic climatic regions (Figure 4.9) are defined by similarities in long-term surface-ocean conditions. The simplest classification separates the surface ocean

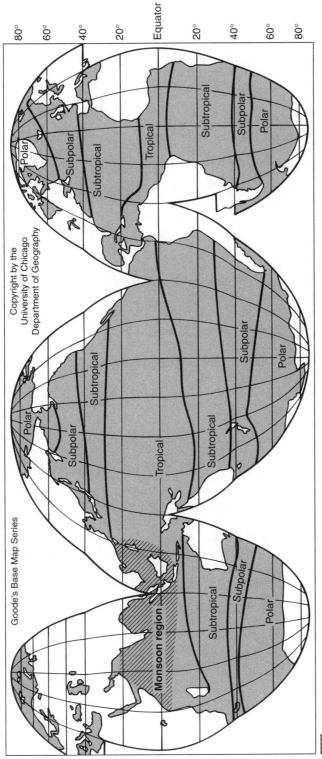

Goode's Base Map Series

Copyright by the
University of Chicago
Department of Geography

▨ Monsoon region, seasonal reversals of winds and current directions.

▨ Open-ocean climatic regions.

FIGURE 4.9
Open-ocean climatic regions.

into coastal (or near-shore) ocean and open ocean. Note that large areas of the northern Indian Ocean are affected by the **monsoons** (seasonally changeable winds) that dominate Southeast Asia (Figure 4.9).

Coastal oceans (discussed in Chapter 8), generally lying over continental shelves, are highly variable. They are influenced by nearness to continents, river discharges, bottom topography, and complex shorelines. In contrast, surface waters in the open ocean are more uniform in temperature and salinity. In the open ocean far from land, climatic zones extend nearly east-west across the ocean (Figure 4.9).

Boundaries separating open-ocean climatic zones are somewhat arbitrary; those separating climatic regions are usually marked by **convergences** (except in the tropics), where surface waters flow to certain areas and sink below the surface. Oceanic convergences correspond to atmospheric weather fronts, but they do not change as rapidly.

Tropical regions straddle the equator in the Atlantic and Pacific oceans (Figure 4.9), but in the Indian Ocean, the tropics lie mostly south of the equator. In the tropics, seasonal temperature changes are slight, and are often less than daily changes. Precipitation near the equator is much larger than evaporation; elsewhere in the tropics, evaporation dominates. Weak and variable winds cause little mixing of surface waters, and the pycnocline is usually shallow.

Subtropical regions, centered around 30°N and 30°S, lie on either side of the tropics, and both north and south of the equator, the Trade Winds continually blow. Because of the strong prevailing winds and abundant sunshine, evaporation exceeds precipitation (Figure 4.8).

Seasonal temperature changes are relatively large in subtropical regions, ranging between 6 and 18°C (43 and 64°F) in open surface waters, but are greatest in enclosed basins, such as the Black Sea. Because of evaporation, subtropical surface waters have above-average salinities and temperatures. When they cool during winter, the high salinities result in increased water density, which causes convective mixing, increasing the depths of regional thermoclines and pycnoclines.

Subpolar regions have an excess of precipitation and lie in a belt of strong winds, especially in the Southern Hemisphere ("Roaring Forties"), where there is little land to obstruct the winds. During seasons of high rainfall or large river discharges, well-developed haloclines form. During local summers, thermoclines develop, but disappear during winter.

In coastal oceans, river runoff reduces local surface salinities. For example, surface salinities in the coastal subpolar North Pacific Ocean are less than 32. (We discuss this further in Chapter 8.)

Polar regions are influenced by winter freezing and summer thawing of sea ice. When sea ice forms, highly saline brines are released, which mix with nearby waters; thus, convective mixing of the surface zone usually accompanies sea-ice formation. Because of prevailing low temperatures, there is little evaporation.

Near Antarctica, chilling of relatively high-salinity Atlantic waters increases seawater density substantially. Chilled surface waters mixing with cold, highly saline brines released from freezing sea ice form Antarctic Bottom Water, the densest water mass in the ocean. These dense waters sink to the bottom, and after flowing around Antarctica, they spread northward, flowing along the bottom into all three major ocean basins. There will be more about this in Chapter 5.

4.7   GREENHOUSE EFFECT

One factor possibly causing the apparent warming trend is the changing composition of Earth's atmosphere. About half the carbon dioxide released by burning fossil fuels since 1850 has collected in the atmosphere. (The fate of the other half is still hotly debated.) This increase in atmospheric carbon dioxide concentration causes increased heat retention in the atmosphere (Figure 4.10).

Consequently, atmospheric carbon dioxide concentrations have risen about 25% above their pre-Industrial Revolution levels. This in turn has apparently increased the **Greenhouse Effect** (warming of Earth's climate similar to the

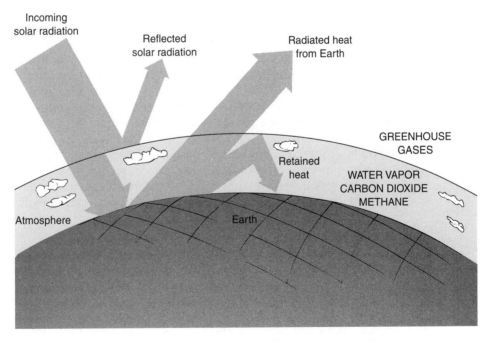

FIGURE 4.10
Earth's radiation balances may be altered by increased atmospheric concentrations of carbon dioxide. This is called the Greenhouse Effect and may cause global warming.

solar warming that occurs in greenhouses). As carbon dioxide levels continue to rise, Earth's surface temperatures are expected to increase with them. Predictions of the amount of warming range from +2.5 to +8°C (+4.5 to 14°F) over the next 50 years. The effect on the ocean will be most noticeable in the Arctic, where sea ice will likely be affected. Indeed, the permanent ice cover in the Arctic Ocean may even disappear. Sea level is expected to rise globally (except near Antarctica), flooding low-lying coastal regions. Such problems will likely be especially acute in small, low-lying, open-ocean islands.

4.8 EL NIÑO–SOUTHERN OSCILLATION (ENSO)

Weather worldwide is affected by interactions between ocean and atmosphere. The most striking example is **El Niño,** the irregular (3 to 7 years) appearance around Christmas of anomalously warm surface waters off Peru and Ecuador (Figure 4.11). The principal cause is a shift in the atmosphere, called the **Southern Oscillation.** The causes of this shift are still disputed, but its effects on the ocean and on climate worldwide are well known.

Equatorial ocean-surface waters are quite warm. This surface layer is about 100 m (300 ft) thick off South America but about 300 m (1000 ft) thick in the western Pacific near Asia (Figure 4.12). This wedge-shaped surface zone is maintained by the Trade Winds, which blow warm surface waters westward near the equator. These same winds normally cause upwelling of cold subsurface waters off South America, which accounts for the usual thinness of the surface zone on the eastern side of the basin. These cold surface waters also cause desert climates in nearby coastal areas.

Near Asia, the warm, surface waters accumulate, forming the thick part of the surface wedge along the equator. (In other words, the equatorial surface currents cannot transport warm waters eastward as fast as they accumulate north of Australia, around Indonesia.)

The presence of such warm waters causes unusually large numbers of thunderstorms in the western Pacific. As warm waters continue to accumulate near Indonesia, the storminess increases and usually extra-powerful hurricanes (sometimes twin hurricanes) set in motion an eastward-moving wave affecting the equatorial surface waters across the Pacific.

When this wave reaches South America, it thickens the surface layer of warm waters there. Now upwelled waters come from within the thickened lens of warm, nutrient-poor surface waters rather than from cold, nutrient-rich subsurface waters. (Remember, upwelling draws waters from 100 to 200 m, or 300 to 700 ft, below the surface.) These warm surface waters cause high rainfall and flooding along the normally arid coast of South America. Elsewhere the atmospheric circulation changes, causing droughts as far away as Australia and northeast Brazil and increased rainfall in Florida and in the equatorial islands of the central Pacific (Figure 4.13).

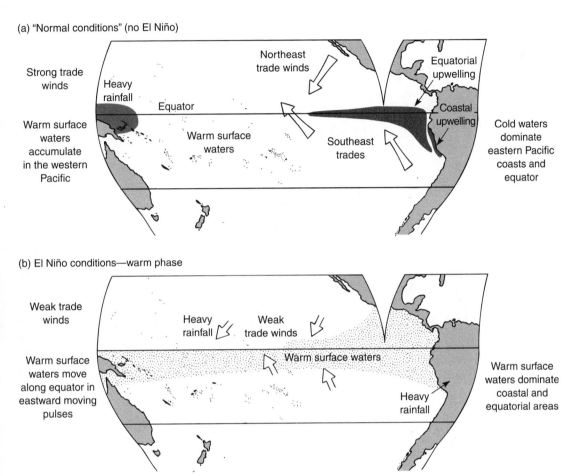

FIGURE 4.11

(a) Surface ocean conditions in years with no El Niño ("normal conditions"). (b) El Niño conditions when Trade Winds are weak. Note that the high-rainfall area has moved from its "normal" position, north of Australia.

Reducing nutrient supplies to the surface waters also causes food production by phytoplankton to be substantially reduced. (These processes are discussed in Chapter 7.) This in turn reduces availability of food for marine animals. Fishing collapses, because there are fewer fish available and any survivors move to find colder waters either offshore or deeper, putting them beyond the reach of nets used by local fishermen. Fish-eating birds also begin to starve, causing them to abandon their eggs and young as they leave to seek food for their own survival. El Niño occurrences in 1968–69, 1972–73, and 1982 decimated the Peruvian anchovy fishery, once the world's largest. El Niño effects on

(a) "Normal"

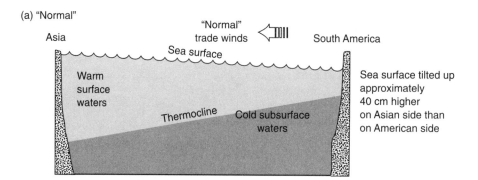

(b) El Niño—warm phase

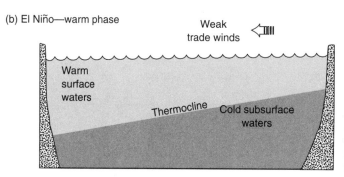

(c) La Niña—cold phase

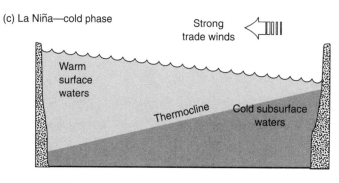

FIGURE 4.12

Changes in the Trade Winds shown in Figure 4.11 also cause changes in the thickness of the layer of warm surface waters that occurs along the Pacific equatorial ocean. (a) During a "normal" year (no El Niño), the wedge of warm waters is thickest near Asia and thinnest near South America. (b) During El Niño conditions, the warm surface layer thins near Asia but thickens substantially near South America. (c) During the cold phase of El Niño (called La Niña), stronger-than-normal Trade Winds cause an unusually thin layer of warm water in the eastern equatorial Pacific, leading to colder-than-normal waters upwelling along the coast and along the equator. (Simplified after K. Wyrtki)

marine life are felt along the coast of the Americas as far away as Alaska on the north and Chile on the south.

El Niño events also influence weather worldwide. During the exceptionally strong El Niño event of 1982, the California coast was battered by unusually severe winter storms. Prolonged droughts and unusually rainy seasons were felt worldwide (Figure 4.13). The 1990–1994 El Niño also influenced weather for an unusually long time.

After an El Niño event ends, the tropical ocean reverts to its more normal state. The surface zone off South America thins so that cold waters again upwell along the equator and off the coast of South America. (This condition has been called **La Niña,** or the cold phase of the cycle.) The western Pacific again begins to accumulate warm surface waters, and the surface zone thickens there. (The rate at which these warm waters accumulate may be a prime factor in determining the frequency of El Niño events.)

Sometimes, the equatorial system rebounds to an unusual extent. After the El Niño event of 1987, a very large band of cold water formed along the equator,

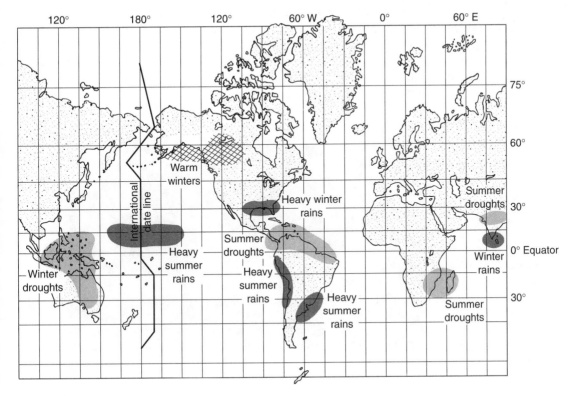

FIGURE 4.13
Precipitation and temperatures are changed worldwide during El Niño events.

which altered the Atlantic hurricane season and caused the jet stream to shift northward over North America. This shift in jet stream location apparently contributed to the North American drought and heat wave of 1988, the worst since the 1930s. Understanding these long-range connections (called **teleconnections**) between ocean and atmosphere will permit better long-range weather and seasonal climate predictions.

Long-term predictions of El Niño occurrences will greatly benefit agriculture and forestry around the world. Even the primitive predictions now made have improved agriculture in Peru. During El Niño years, crops needing water, such as rice, are planted. In non-El Niño years, cotton and other dry-weather crops are planted. This has stabilized farming incomes, which were previously decimated by the El Niño events. Similar results have been obtained in parts of southern Africa. Agriculture in the southeastern United States would save many millions of dollars per year if accurate El Niño forecasts were routinely available.

4.9   ICEBERGS

Glaciers on land discharge ice into the ocean. **Icebergs** come from two different sources. Glaciers draining mountains discharge irregularly shaped bergs (Figure 4.14). The large floating ice shelves around Antarctica form large flat-topped bergs, some of them as large as small states (Figure 4.15). Icebergs are moved by currents and, to a lesser degree, by winds. Larger ones can persist for years before melting.

FIGURE 4.14
Small, irregularly shaped icebergs break off mountain glaciers when they discharge into the ocean. Such a small iceberg will normally melt within a year after entering the ocean. (Courtesy NASA)

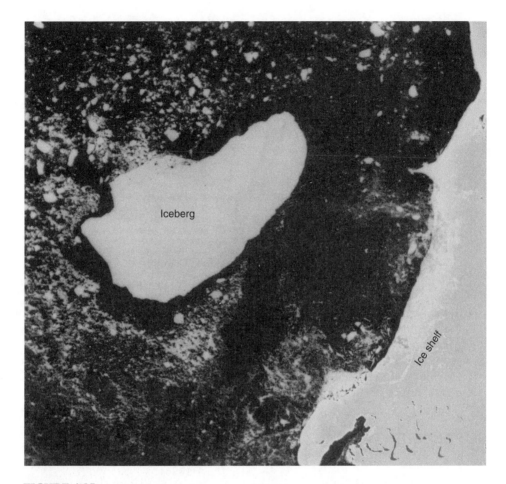

FIGURE 4.15
Gigantic tabular icebergs break off the floating ice shelves around Antarctica. This one, formed in late 1976, is roughly the size of the state of Rhode Island and will persist for decades before it finally melts.
(Courtesy NASA)

4.10   MONITORING THE OCEAN

Increased concerns about changes in Earth's climate are leading to continuing efforts to detect changes in the open ocean, which are the long-term memory for the atmosphere—a critical part of Earth's climatic system. Satellites are an important part of the ocean-monitoring system.

Earth-orbiting satellites collect data that permit monthly and annual maps of surface-water temperatures and distributions of other properties to be made routinely. Such maps were impossible when ships were the only platforms available for studying the ocean.

Ships will also be used to measure water temperatures and other properties in areas of special importance, such as areas of bottom-water formation. Such observations can provide information about possible changes in the deep-ocean waters to supplement the data on the ocean surface available from satellites. The behavior of sound in seawater is also used to study the open ocean's interior over larger distances than a ship can cover.

QUESTIONS

1. What are the average temperature and average salinity of the ocean?
2. Explain how the ocean is heated primarily in the low latitudes and cooled in the high latitudes.
3. How is heat transported from the tropics to the high latitudes?
4. List the three processes that change seawater salinity. Where is each most important?
5. List and briefly describe the three oceanic depth zones.
6. What three processes cause heat loss from the ocean's surface? Which process has the greatest effect on the atmosphere? Why?
7. What processes cause seasonal variations in surface-water properties? Which are most important in the subtropical ocean? Which are most important in the polar ocean?
8. How are oceanic climatic zones related to atmospheric circulation?

SUPPLEMENTARY READINGS

Books

Miller, A., and Anthes, R. A. *Meteorology.* 6th ed. Columbus, OH: Merrill Publishing Company, 1992. General reference, elementary.

Perry, A. H., and Walker, J. M. *The Ocean-Atmosphere System.* London: Longman, 1977. Emphasizes ocean-atmosphere interactions.

Articles

Gregg, M. "Microstructure of the Ocean." *Scientific American* 228(2):64–77.

MacIntyre, Ferran. "The Top Millimeter of the Ocean." *Scientific American* 230(5):62–77.

Stewart, R. W. "The Atmosphere and the Ocean." *Scientific American* 221(3):76–105.

KEY TERMS AND CONCEPTS

Light and heat absorption
Insolation
Layered structure
Surface zone
Mixed layer
Thermocline

Pycnocline
Pycnocline zone
Deep zone
Water masses
Heat budget
Evaporation

Precipitation
Halocline
Oceanic climatic zones
Monsoons
Tropical regions
Subtropical regions
Subpolar regions
Polar regions
Greenhouse effect

El Niño
Southern Oscillation
Trade Winds
La Niña
Teleconnections
Icebergs
Ocean monitoring
Satellites
Global ocean observations

5

Ocean Currents and Climate

Ocean waters move unceasingly. Anyone who sails or swims in the ocean experiences **currents** (horizontal water movements). Some currents affect only small areas, such as beaches; these are the ocean's response to local conditions. Other currents extend over large ocean areas and transport heat from the tropics, where Earth is warmed, to the polar regions, where it is cooled, and then back again to the tropical ocean. In this chapter, we learn about the causes and consequences of ocean currents, both at the surface and in the ocean's interior.

5.1 SURFACE CURRENTS

Our knowledge of ocean-surface currents and winds comes primarily from compilations of sailor's observations. The American colonial postmaster, Benjamin Franklin (1706–1790), synthesized the knowledge of his cousin, a Nantucket whaler, and in 1770 published the first map of the Gulf Stream. Sailors knowing about this strong current could cut weeks off their westward passages.

Our present knowledge of surface currents began with the work of a crippled American sailor. After an accident ended his seagoing days, Matthew Fontaine Maury (1806–1873) compiled weather and ocean observations from ship's log books, collected by the U.S. Navy. From these he published a chart of average surface winds in 1840. (He was also instrumental in establishing the international observing system that made weather forecasting possible.)

Maury estimated currents by analyzing recorded deflections of ships' courses, which were caused by surface currents. A current deflects a ship's course so that its actual final position differs from its intended destination. Combining thousands of such observations, Maury compiled maps of average ocean currents (Figure 5.1); these data still provide our global picture of surface currents, although satellite observations are showing much greater variability than could be detected by the scattered observations available from ship's logs.

Seasonally changeable currents, such as the Monsoon Currents in the Northern Indian Ocean, were not clearly delineated by such averaged observations. Understanding changeable currents requires repeated observations within a single season. Because it is difficult to map transient currents, we are still learning about the ocean's variability, primarily from satellite observations. (More about this later.)

Surface winds and surface currents are closely linked. You can see this by comparing the patterns of prevailing winds with the ocean-surface current patterns (Figure 5.2). Patterns of prevailing winds and average currents correspond closely, except around Antarctica.

5.2   GYRES

Open-ocean surface currents form nearly closed current systems, called **gyres.** Each large basin has a current gyre in its subtropical regions (around 30°N and 30°S). Smaller gyres occur in the subpolar oceans, centered near 50°N. In the Southern Ocean, the **Antarctic Circumpolar Current** flows around Antarctica, connecting the gyres in all three basins. Each open-ocean gyre consists of four currents. East-west currents form the northern and southern portions; these, in turn, are connected by generally north-south currents flowing along continental margins.

Open-ocean currents, such as the **North Pacific Current** or the **North** and **South Equatorial currents,** move slowly, at speeds of 3 to 6 km/day (2 to 4 mi/day), and are generally shallow, usually extending 100 to 200 m (300 to 700 ft) below the surface. Waters moving in these currents remain in the same climatic zone for months, giving them time to equilibrate with local conditions.

The ocean's largest current is the eastward-flowing Antarctic Circumpolar Current, which circles Antarctica and connects southern basin gyres. Such a globe-encircling current is possible only in the Southern Ocean around Antarctica; elsewhere, continents deflect east-west currents. The Drake Passage, the narrow strait between Cape Horn (the tip of South America) and the Antarctic Peninsula, deflects waters from the Antarctic Circumpolar Current to form the northward-flowing Peru Current along the west coast of South America.

Western boundary currents are the strongest currents in each gyre; they flow northward in the Northern Hemisphere and southward in the Southern Hemisphere (Figure 5.1). Prominent among them are the **Gulf Stream System** (North Atlantic) and the **Kuroshio** (off Japan), which, next to the Antarctic Circumpolar Current, are the ocean's strongest currents.

Western boundary currents move water between 40 and 120 km (25 and 75 mi) per day. Their flows also extend much deeper than other currents, down

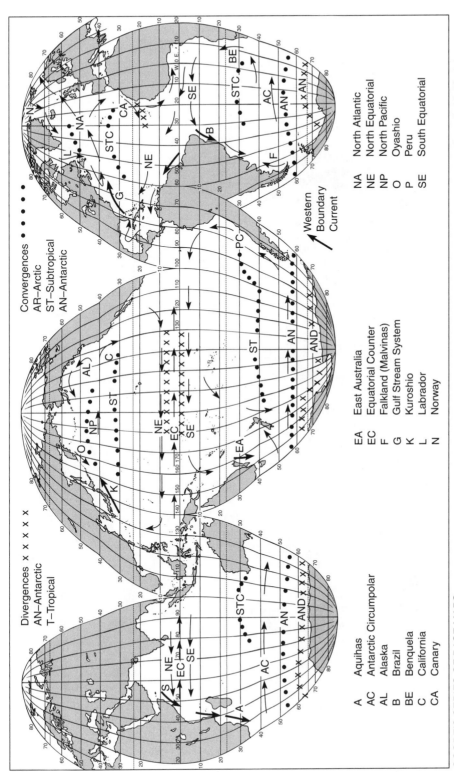

GOODE'S SERIES OF BASE MAPS
HENRY M. LEPPARD, EDITOR

Prepared by J. Paul Goode
Published by the University of Chicago Press, Chicago, Illinois
Copyright 1917 by the University of Chicago

FIGURE 5.1

Ocean-surface currents during Northern Hemisphere winter.

93

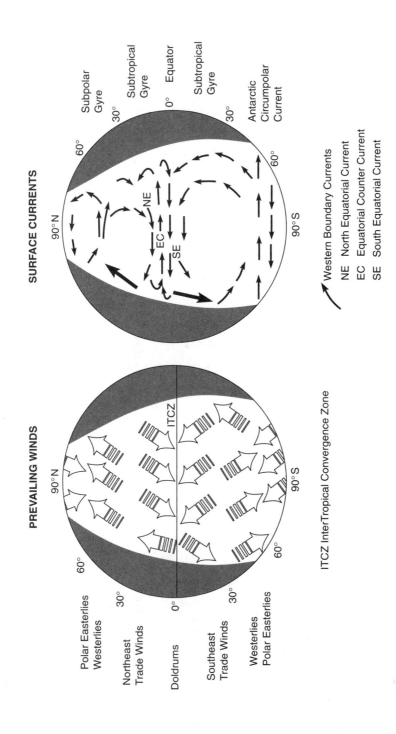

FIGURE 5.2

Prevailing winds and generalized surface currents.

[After R. H. Fleming, "General Features of the Ocean," *Geological Society of America Memoir 67(1)* (1957), p. 95]

to depths of 1000 m (3000 ft) or more. Unlike the surface currents, these **western boundary currents** are affected by ocean-bottom features and erode bottom sediment deposits. The flow of each of these currents is 50 to 100 times the total discharge of all the world's rivers.

Western boundary currents transport large amounts of heat from the tropics toward the poles. Heat transported by the Gulf Stream System helps to keep Northern Europe warmer than comparable latitudes in North America. Western boundary currents of the Southern Hemisphere, such as the **Agulhas Current** (off southeast Africa), the **Brazil Current,** and the **East Australia Current,** are weaker than those in northern oceans. In part, the relative weakness of the Southern Hemisphere western boundary currents is caused by the scarcity of land there. In the absence of barriers, western boundary currents cannot form.

Eastern boundary currents, such as the **California Current** and the **Canary Current** (off west Africa), are slower, shallower, and broader than their western counterparts. In fact, most currents in the gyres are fairly shallow and slow moving.

Transport of colder waters toward the tropics in eastern boundary currents is another way in which heat is transported. These currents are comparable to the return flows in circulating water systems of household heating systems. These cold waters are heated in the tropics and then transport heat poleward.

5.3   TRADE WINDS

Trade Winds (Figure 5.2) straddle the equator on the north and south. These persistent winds drive both **North** and **South Equatorial currents** westward, thus transporting warm surface waters westward. **Equatorial Counter Currents** return these warm waters eastward, along the equator. (We discussed this in Chapter 4 when we considered El Niño events.)

The Southeast Trade Winds of the Southern Hemisphere extend across the equator into the Northern Hemisphere. The Doldrums (a region of light and variable winds, also called the **Intertropical Convergence Zone** or ITCZ) separate the Trade Wind systems of the two hemispheres. The ITCZ changes position seasonally, lying generally north of the equator—especially in the Atlantic, where it generally stays north of the equator. Consequently, the eastward-flowing Equatorial Counter Current, separating the surface current systems of the two hemispheres, also lies north of the equator.

The South Equatorial Current crosses the equator in the Atlantic and to a lesser extent in the Pacific. In this way, it transports surface waters into the Northern Hemisphere. The return flow is through subsurface currents (which we discuss later).

5.4 CORIOLIS EFFECT

Surface waters, ice, and anything else set in motion by winds move obliquely to the right of the wind in the Northern Hemisphere, and to the left of the wind in the Southern Hemisphere. Caused by Earth's rotation from west to east, this deflection, called the **Coriolis effect,** of objects moving over Earth's surface causes currents and winds to apparently change directions as seen from Earth's surface (Figure 5.3), but not if seen from the Moon, as we shall now explain.

Let us see what is happening. A bit of air or water (it will be called a particle here) moving due northward from the equator also is moving eastward at about 1670 km/hr (about 1040 mi/hr) because of Earth's eastward rotation. After moving northward to latitude 30°N (about the latitude of New Orleans), the same particle (not attached to Earth's surface) is still moving eastward at 1670 km/hr. But at New Orleans, where Earth's surface moves eastward more slowly, at about 1450 km/hr (900 mi/hr), the particle is apparently moving eastward faster than Earth's surface at that point. To a ground-based observer at New Orleans watching the particle, it appears to have been deflected eastward (to the right of its original path).

A Moon-based observer, however, would see that the particle was not actually deflected but continued moving northward in a straight line. But points on Earth's surface at different latitudes moved eastward at speeds that increased closer to the pole, causing the apparent deflection.

FIGURE 5.3
Paths of moving objects are deflected by the Coriolis effect. There is no deflection at the equator, and the amount of deflection increases toward the north and south poles.
[After A. N. Strahler, *Physical Geography,* 2d ed. (New York: John Wiley & Sons, Inc., 1960)]

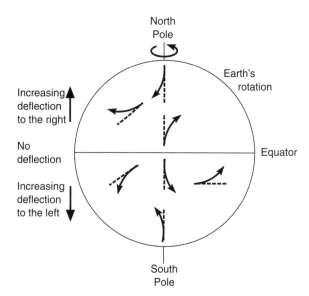

------- Actual paths, seen by Moon-based observer

Deflected path, seen by Earth-based observer

This apparent deflection of moving particles—to the right in the Northern Hemisphere and to the left in the Southern Hemisphere—is called the Coriolis effect, named for the French scientist Gaspard Gustave de Coriolis (1792–1843), who first explained these deflections as a result of our observing winds and currents from Earth's moving surface.

The amount of deflection depends on the latitude and the particle's speed. There is no deflection of particles moving along the equator; apparent deflections are greatest at the poles. Resting particles are unaffected. The Coriolis effect is important when other forces acting on a moving particle are small and when the particle has moved a long distance. (Your car is not affected by the Coriolis effect because it is in contact with the ground.)

5.5   EKMAN SPIRAL

Winds blowing steadily across water move it by dragging on the surface. Wind ripples or waves cause the surface roughness necessary for the wind to "grab" surface waters and set them in motion. A steady wind blowing for 12 hours at an average speed of 100 cm/sec (about 2.2 mi/hr) over deep water causes surface currents of approximately 2 cm/sec, about 2% of the speed of the wind.

Although winds initially set surface waters in motion, the resulting currents commonly extend down to about 100 m (300 ft) below the surface. This is caused by slowly moving fluids flowing as thin sheets over each other (called **laminar flow**). Because of water's viscosity, each moving layer drags on the layer below it. At these low speeds, momentum is transmitted from one layer to another by collisions of individual water molecules. Momentum is a combination of how big the objects are, how fast they are going, and their direction of motion.

As energy transfers from rapidly moving layers to more slowly moving ones, it is steadily lost in overcoming molecular viscosity (caused by molecular interactions) within layers. Consequently, wind-induced surface currents do not extend downward to great depths in the ocean. Remember also that in the ocean, the pycnocline also inhibits downward energy transfers.

Ocean waters commonly move too rapidly for laminar flow to persist. Instead, currents are usually **turbulent** (particles moving in irregular, ever-changing eddies, carried along by the main flows). These eddies interact, transferring motion from one to another. Resistance to flow resulting from eddy interactions is called eddy viscosity. Eddies transfer momentum many thousand times more rapidly than energy is transferred between molecules in laminar flows. Eddy viscosity varies greatly, depending on density stratification and flow speeds. Now, let's put all these pieces together.

Steady winds blowing across an infinite, homogeneous ocean with uniform eddy viscosity in the Northern Hemisphere (remember that this is an idealized case) cause surface waters to move at an angle of 45° to the right of the wind (45° to the left in the Southern Hemisphere). Each layer sets the layer below in

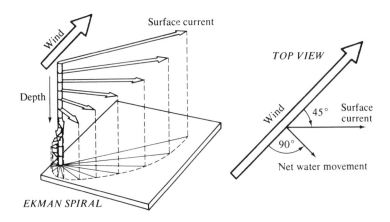

FIGURE 5.4
Water movements in wind-generated currents in the Northern Hemisphere. The sense of
the current deflections is reversed in the Southern Hemisphere; thus, currents are
deflected to the left of the winds that set them in motion.

motion. As each deeper layer is set in motion, it too is deflected by the Coriolis
effect, so that it moves farther to the right of the overlying layer. Deeper layers
move more slowly because energy is lost in each transfer between layers.

If we plot the movement of each layer using an arrow (called a vector)
whose length represents the speed of movement and whose direction corresponds to the direction of that layer's movement, we see an idealized pattern of
surface currents set in motion by Northern Hemisphere winds (Figure 5.4). This
is called an **Ekman Spiral,** named after the Swedish physicist V. W. Ekman
(1874–1954), who first explained this relationship between winds and surface
currents. As each moving layer is deflected to the right of the overlying layer,
directions of water movements also shift with increasing depth. Eventually,
often at a depth of about 100 meters (about 300 ft), water is moving slowly in a
direction opposite to that of the surface layer. This is considered to be the base of
wind-driven currents.

Combining movements in all layers in the Ekman Spiral, we find that **net
water movement** is perpendicular to wind direction (Figure 5.4), 90° to the
right of the wind in the Northern Hemisphere (90° to the left in the Southern
Hemisphere). This is a simplified model for an infinite, homogeneous ocean with
no pycnocline and no boundaries. This is a typical example of how such problems are attacked. Complicating details of the actual situation are omitted from
idealized models until the situation is sufficiently simple that it can be solved.
Then the problem is to interpret the actual ocean, using the idealized results.

Because the ocean does not satisfy such idealized conditions, actual wind-
induced water movements can differ appreciably from theoretical predictions.
For example, the angle between wind direction and surface-water movement
varies from 15° in shallow waters to the theoretical maximum of 45° in deep

waters. The pycnocline also inhibits energy transfers to deeper waters, in most cases confining the wind-driven currents to the surface zone. The pycnocline is effectively the floor for surface currents.

5.6   GEOSTROPHIC CURRENTS

Now we turn to large surface-current systems caused by prevailing winds (Figure 5.2). In deep waters, net water movements (in the Northern Hemisphere) are 90° to the right of the prevailing winds (Figure 5.4), so Earth's prevailing winds move surface waters toward the center of each subtropical gyre (Figure 5.5); this is called a **convergence.** These wind-induced convergences form low hills (Figure 5.5) of less dense surface waters near the centers of gyres, because prevailing winds generally blow around the basin margins. The resulting sea surface slopes are very gentle, only 1 to 2 m or 3 to 7 ft (usually less) above a reference level well below the surface, and extend over thousands of kilometers. In contrast, the subpolar gyres cause **divergence** in the center of the gyre. As we shall see later, this influences the amount of deeper waters that can come to the surface.

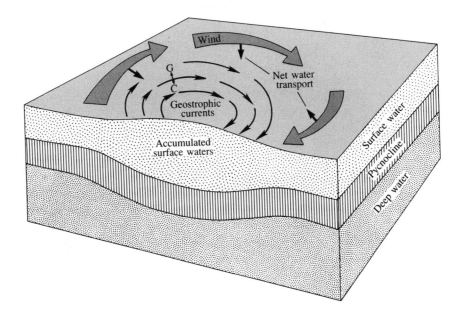

FIGURE 5.5

Net water movements resulting from prevailing winds in the Northern Hemisphere sub-tropical gyre cause surface waters to accumulate (a convergence zone), forming a low hill of water and depressing the pycnocline. Geostrophic currents result from waters flowing down the sloping sea surfaces until they achieve a balance between these gravity effects and the deflection (to the right in the Northern Hemisphere) caused by the Coriolis effect.

The steepest slopes occur on the western sides of ocean basins, where currents are strongest, but even there such slopes can rarely be measured directly, as we might survey a hill on land. (It is possible to measure the slopes of the Gulf Stream where it passes between Florida and the Bahamas; the difference in elevation is about 1 m there.) Instead, oceanographers calculate "dynamic topography" using measurements of temperature and salinity. Let's see how they do it.

Remember that a mass of less dense water occupies more volume than a comparable mass of denser water. Furthermore, the density of water is controlled primarily by its temperature and salinity. This means that the water surfaces above two columns of water of equal mass but having different densities will stand at different heights above a chosen level surface below them; the less dense water column will stand higher than the denser water column. Knowing the densities of various columns of water above a chosen deep reference level, we can calculate water-surface slopes between stations. After making many calculations of height above the reference surface, we can map sea-surface slopes over large areas. From this calculated dynamic surface topography, we can calculate speeds and directions of surface currents.

Let's see how this works. Water at the top of the hill responds to a sloping ocean surface by flowing downhill, just as it would on land. The water starts to move downhill, but in the ocean its path is deflected by the Coriolis effect. The water flow changes direction until the force of gravity is balanced by the Coriolis effect. In an idealized, frictionless ocean, gravity acting in a downhill direction is balanced by the Coriolis effect acting to the right (uphill in an equilibrium situation), resulting in a geostrophic (Earth-turned) current. In the Northern Hemisphere, the hill of less dense water is on the right when one looks in the direction of flow; in the Southern Hemisphere, it is on the left (Figure 5.5).

Major ocean currents are **geostrophic.** Thus, these currents can be mapped by charting sea-surface topography. Currents flow around elevations or depressions; slope steepness controls current speed. Current speeds are greater on steep slopes than on gentle slopes.

In an idealized ocean, where water had no viscosity, currents would flow at constant elevations around the hill of water and therefore never reach the bottom. However, seawater has viscosity (resistance to flow), and energy must be expended to keep it flowing downhill. **Geostrophic currents** are simplified representations of a more complex world (like the Ekman Spiral), but they permit useful predictions of tracks of icebergs moved by currents or floating debris from wrecked ships or aircraft.

Satellites now measure directly the subtle topography of the sea surface. Thus, satellites can provide maps of the ocean topography from which currents can be deduced without the need for the many measurements of water temperature and salinity that previously took years to acquire. In short, satellites make it possible to study the changeable current regime, rather than depend on averaged observations taken over many years and often many thousands of kilometers apart.

5.7   GULF STREAM RINGS

The ocean is much more dynamic and changeable than is indicated by maps of average conditions, unlike Franklin's vision of currents as "rivers in the sea," which implied fixed locations. This image of oceanic constancy has been replaced by one of slowly changing ocean currents responding to variations in the winds, freshwater inputs, and solar heating. Oceanographers are still working out the details of these interactions and the resulting variable currents. It is possible to predict such current locations and speeds for small regions for which good observations are available. Soon it will be possible to have frequently updated maps of ocean currents worldwide.

Perhaps the most dynamic and changeable currents are the large, current-bounded **rings** spun off western boundary currents, especially by the Gulf Stream, the Kuroshio, and the Eastern Australian Current. Such rings begin as large **meanders** in the boundary currents, which eventually break off, and the rings then move independently of the currents that formed them (Figure 5.6). Each ring is bounded by strong currents, isolating waters and organisms inside the rings from those outside. Rings are most common in the western portions of ocean basins (Figure 5.7), because they are formed by western boundary currents.

Rings form on both sides of the Gulf Stream (the most extensively studied system). Rings formed on the north side are 100 to 200 km (60 to120 mi) wide and enclose parcels of warm Sargasso Sea waters from south of the Gulf Stream. These are called **warm core rings.** Because the rings are up to 2 km (1 mi) deep, they do not come up onto continental shelves, although they do come close enough to influence shelf currents and distributions of organisms.

On the south of the Gulf Stream, rings contain cold, low-salinity waters from the coastal ocean; they are called **cold core rings.** Rings on both sides of the Gulf Stream move slowly (5 to 6 km/day or 3 to 5 mi/day) southwestward. Warm core rings are reabsorbed into the Gulf Stream after a few months to a year when the rings reach Cape Hatteras, North Carolina. The rings are prevented from flowing farther south by the Gulf Stream flowing along the continental margin. In contrast, cold core rings can last for several years. These ring systems are similar to storms in the atmosphere, but are smaller and longer lived.

5.8   WIND-INDUCED VERTICAL WATER MOVEMENTS

Winds blowing across ocean surfaces also cause vertical water movements, both upward (**upwelling**) and downward (**sinking**). Coastal upwelling and sinking occur where prevailing winds blow parallel to the coast. Winds move surface waters, but shorelines and shallow bottoms restrict lateral water movements (Figure 5.8). When surface waters are moved offshore, they are replaced by relatively shallow waters flowing to the surface, from 100 to 200 m (300 to 700 ft)

FIGURE 5.6
Rings form from meanders in western boundary currents. After separating, rings move slowly southwestward (opposite to the boundary current).

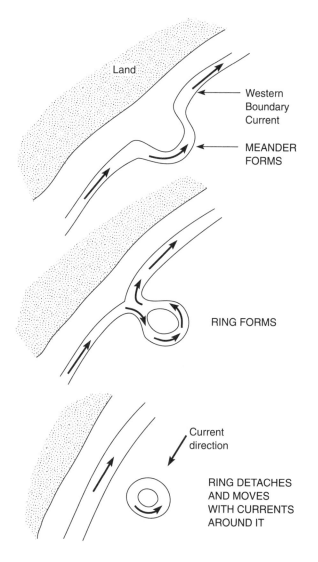

deep. These waters often come from below the pycnocline, and are thus usually colder than surface waters.

Coastal upwelling occurs along the western coasts of continents. Cool summer weather with coastal fogs is caused by the presence of cooler, upwelled waters. Upwelled waters are also identifiable by low dissolved-oxygen contents, because they have long been out of contact with the atmosphere. Vertical water movements also bring waters rich in dissolved nutrients (nitrogen and phosphate compounds) to the surface, where they support phytoplankton growth.

Thus, many upwelling areas support major fisheries; indeed, about half the world's catch comes from upwelling areas.

Equatorial upwelling also occurs. In this case, wind-induced upwelling is caused by the Coriolis effect changing direction at the equator (acting to the right north of the equator, but to the left south of the equator). Westward-flowing, wind-driven surface currents near the equator flow northward on the north side of the equator and southward on the south side; this is an example of a divergence, where surface waters are moved away from the location and replaced by upwelling deeper waters.

Downward movements (sinking) of coastal waters result from winds moving surface waters shoreward. Coastal sinking is less obvious to coastal dwellers than upwelling, although the abundances and distributions of fishes may be changed radically by sinking surface waters. Such transitory shifts are especially noticeable to swimmers. Nearshore surface waters are warm while sinking is occurring but much colder during upwelling.

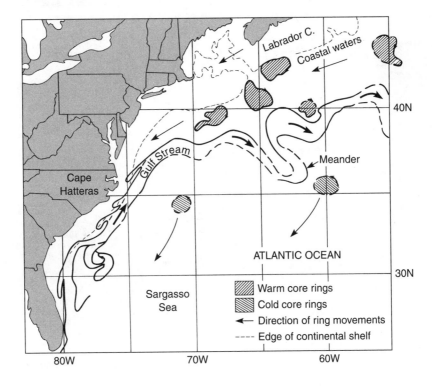

FIGURE 5.7
Rings are common along the U.S. Atlantic coast. Warm core rings enclose waters from the warmer Atlantic Ocean. Cold core rings contain colder, coastal waters.

FIGURE 5.8
Winds cause coastal upwelling and sinking in the Northern Hemisphere. Sea-surface and pycnocline slopes are greatly exaggerated. The arrows show directions of water movements.

UPWELLING

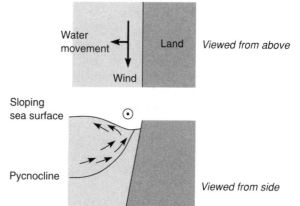

SINKING

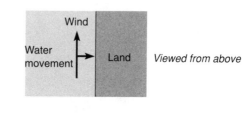

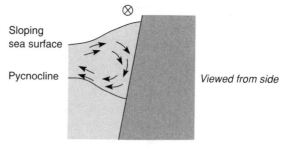

⊙ Wind blowing toward you

⊗ Wind blowing away from you

5.9 WATER-MASS MOVEMENTS

Deep-ocean water masses form primarily in polar regions, especially near Antarctica and Greenland (Figure 5.9). In both areas, surface waters are chilled and freeze, forming sea ice. Salts excluded during freezing of sea ice mix with the already cold waters. Because of this pronounced cooling and increased salinity, cold, dense waters form and sink below the surface. The dense waters formed near Greenland flow southward in the Atlantic (called the **North Atlantic Deep Water,** NADW). Dense water masses formed near Antarctica (Figure 5.10) flow around the continent before eventually flowing northward along the bottom into the three major basins, forming the **Antarctic Bottom Water** (AABW), the ocean's densest water mass.

Antarctic Intermediate Water (intermediate density) forms near Antarctica. It flows northward at intermediate depths where the slightly denser North Atlantic Deep Water and Antarctic Bottom Water lie below it (Figure 5.11). Eventually these dense waters mix with waters around them and can no longer be separately identified.

Several other water masses, also identified by characteristic temperatures and salinities, form in partially isolated, coastal seas. For instance, warm, saline waters flow out of the Mediterranean Sea through the Strait of Gibraltar; a similar water mass forms in the Red Sea. Both form thin layers of warm, salty waters

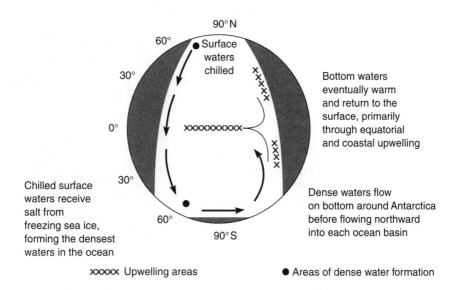

FIGURE 5.9
Movements of bottom waters in the Atlantic Ocean after they form near Greenland in the North Atlantic and are rechilled near Antarctica. The resulting current of dense waters flows eastward along the bottom around Antarctica. Eventually these bottom waters flow northward into all the major ocean basins.

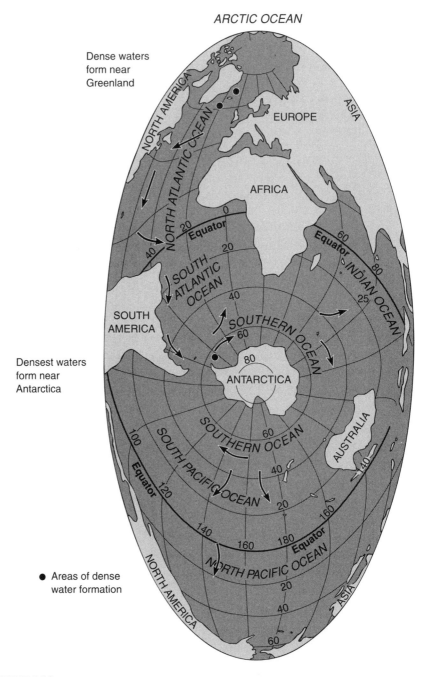

FIGURE 5.10
General movements of bottom waters in all ocean basins, at depths greater than 4 km (2.4 mi).

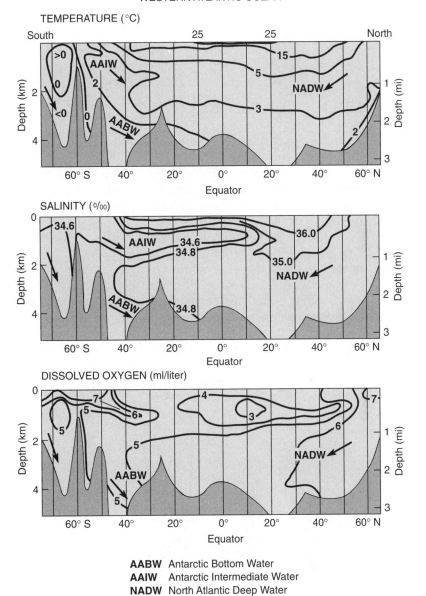

WESTERN ATLANTIC OCEAN

AABW Antarctic Bottom Water
AAIW Antarctic Intermediate Water
NADW North Atlantic Deep Water

FIGURE 5.11
Vertical distributions of water temperatures, salinities, and dissolved-oxygen concentrations in the Atlantic Ocean (After Wüst). Arrows show general water movements.
[After H. U. Sverdrup, M. W. Johnson, and R. H. Fleming, *The Oceans: Their Physics, Chemistry, and General Biology* (Englewood Cliffs, NJ: Prentice-Hall, Inc., 1942), p. 748]

at intermediate depths and can be traced over great distances as they move through the nearby oceans.

5.10   DEEP-OCEAN CURRENTS

Below the pycnocline, intermediate and deep waters move sluggishly through the ocean basins. These deep currents are isolated from wind-driven surface currents by the pycnocline; the deep currents are driven primarily by differences in seawater density, which are controlled by variations in temperature and salinity. Hence, the deep-ocean circulation is called the **thermohaline circulation** (thermo = heat; haline = salt).

Unfortunately, in the deep ocean it is difficult to make direct measurements, and we have nothing like Maury's compilations of average surface currents. Instead, slight changes in water temperature, salinity, and dissolved-oxygen concentrations provide most of our information. Heat, salt, and dissolved gases move slowly across boundaries between adjacent layers through mixing and diffusion, and so these deep-ocean water masses can be identified over large distances.

Little is known about these slow, deep currents because they have been difficult to measure directly (Figure 5.12). Recently, submerged floats have been

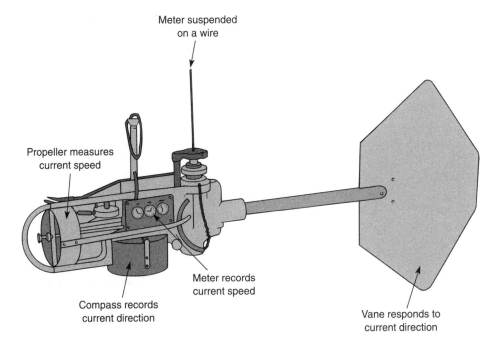

FIGURE 5.12
Current meters measure current directions and speeds. An early mechanical current meter, shown here, works like an anemometer, which measures wind speed and direction. Modern current meters use electrical or acoustic sensors to determine current directions and speeds.

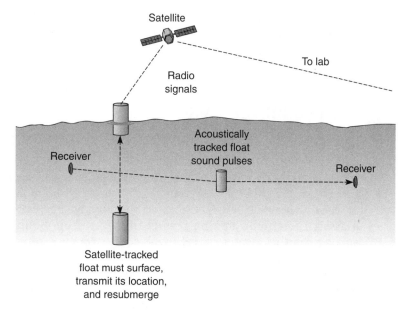

Satellite

To lab

Radio
signals

Receiver

Acoustically
tracked float
sound pulses

Receiver

Satellite-tracked
float must surface,
transmit its location,
and resubmerge

FIGURE 5.13

Deep-ocean water movements are measured by floating instruments, designed to float at preset densities. Their movements are tracked by sound signals emitted by the float and detected by listening stations ashore, which can plot their positions. Other floats can come to the surface, where they communicate their positions to satellites and then sink back to their original depths.

developed, which can be tracked over long distances using sound signals (Figure 5.13) to study movements of individual water parcels, but the movements are so slow and variable that it is difficult to obtain enough data to map these weak currents. Various **tracers,** such as dissolved oxygen, and pollutants, have been used to track movements of these subsurface waters.

Deep-ocean currents move generally north-south (Figure 5.10), except around Antarctica. Unlike surface currents, bottom currents cross the equator. Subsurface currents in the Atlantic connect both northern and southern polar regions. Like surface currents, deep-ocean or bottom currents are strongest on the western sides of basins.

Bottom-water flows are also partially deflected by seafloor mountain ranges. For instance, gaps in the Mid-Atlantic Ridge channel bottom-water flows from the western Atlantic basins across the Mid-Atlantic Ridge into the eastern Atlantic basins. Conversely, the Greenland-Iceland Ridge separating the Arctic Sea from the Atlantic Ocean prevents dense waters formed in the Arctic Ocean from flowing out into the Atlantic. Only water masses formed near Greenland (south of the ridge) can readily flow into the North Atlantic.

Subsurface waters slowly return to the sea surface by slow, upward movements of deep waters. These return flows to the surface appear to be concentrated in equatorial and coastal upwelling zones. An unknown amount of deep water flows up through the pycnocline, warmed by heat coming from the ocean bottom and especially from recently erupted volcanic rocks still cooling on the East Pacific Rise.

5.11   MIXING PROCESSES

Mixing of ocean waters results from their continual movement. Water molecules are constantly moving randomly and are also carried by eddies, which are transported by currents. Such swirling movements transfer heat, salt, and other properties. If a water parcel is small, mixing occurs by molecular motions. If the parcel is somewhat larger, eddies do the mixing. Finally, currents move water masses. In general, mixing results from water motions of the same size or smaller than the water parcel involved. Movements on scales larger than the water parcels involve transport rather than mixing.

Energy for mixing comes from winds, currents, and tides. Vigorous mixing occurs either near the sea surface or near the sea floor. At the sea surface, a nearly isothermal (or isohaline) surface zone, often tens of meters thick, is formed by wind-induced mixing. Density changes resulting from surface cooling cause sinking of denser waters. This is especially important in polar regions of the North Atlantic and Southern oceans.

Near-bottom flows also cause mixing, which is especially noticeable when strong currents flow across sills or ridges. In shallow waters, bottom-associated mixing is especially important. A layer of well-mixed waters often occurs near the ocean floor, called the **benthic boundary layer.**

Near coasts, strong currents also mix waters, as do breakers in surf zones. Fresh water discharged by rivers mixes with seawater in estuaries and near land; winds and tidal currents supply the energy required for mixing.

Below the pycnocline in the open ocean, there is little energy for mixing of waters. In a stable situation, the water below will be denser than any given water parcel; the water above will be less dense. In this situation, a water parcel will spread laterally, along surfaces of constant density, as a thin layer interleaved between other layers. Mixing occurs at interfaces between these layers. Some is caused by different water masses flowing at different speeds and directions. Diffusion of molecules between adjacent layers also causes mixing.

5.12   ACOUSTIC TOMOGRAPHY

As we discussed in Chapter 4, the behavior of sound in the ocean can be used to study ocean structure and processes over large distances. A recently developed means of determining water-mass distributions uses sound rays much as doctors use x-rays to study organs in human bodies through CAT scans. The technique is

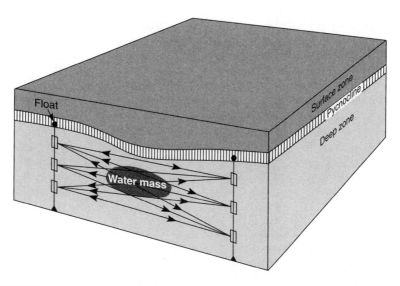

FIGURE 5.14
Acoustic tomography uses multiple sound paths between many sound transmitters and receivers to determine locations and movements of subsurface water masses.

called **acoustic tomography.** Here, entire arrays of sound sources and listening devices are installed around the basin to be studied (Figure 5.14). Sound pulses are repeatedly transmitted by the sound sources and detected by the arrays of listening devices. These signals are then analyzed by computers to detect variations in sound speeds among the many pairs of sources and receivers. From these analyses, it is possible to detect different water masses and to study their movements. Such techniques are especially useful for studying currents in areas where flows are intermittent and too strong for normal current-meter arrays.

5.13   GLOBAL CONVEYER BELT

The ocean's thermohaline circulation has been called a **global conveyer belt** (Figure 5.15). The fundamental idea is that dense waters form near southern Greenland, sink, and flow southward to Antarctica as deep currents (North Atlantic Deep Water), where they are further chilled in Antarctica's Weddell Sea, becoming denser. Afterwards, these dense waters flow into all three major ocean basins, where they gradually warm and return to the surface, and then flow back into the Atlantic to begin the process again. Let's look at the details of this system and their climatic implications for Europe.

The ocean's deep circulation begins with the evaporation of North Atlantic surface waters. The salt remains, causing the North Atlantic surface waters to have the highest average salinity of the three major ocean basins. Water vapor,

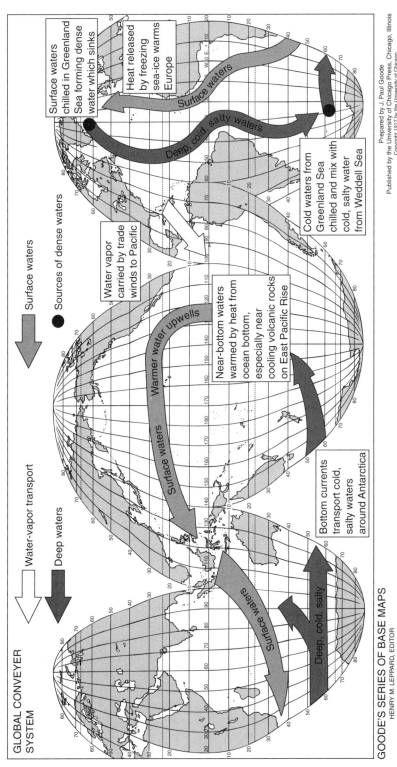

GLOBAL CONVEYER
SYSTEM

Surface waters

Sources of dense waters

Water-vapor transport

Deep waters

Surface waters chilled in Greenland Sea forming dense water which sinks

Heat released by freezing sea-ice warms Europe

Surface waters

Deep, cold, salty waters

Water vapor carried by trade winds to Pacific

Cold waters from Greenland Sea chilled and mix with cold, salty water from Weddell Sea

Warmer water upwells

Surface waters

Near-bottom waters warmed by heat from ocean bottom, especially near cooling volcanic rocks on East Pacific Rise

Surface waters

Deep, cold, salty

Bottom currents transport cold, salty waters around Antarctica

Prepared by J. Paul Goode
Published by the University of Chicago Press, Chicago, Illinois
Copyright 1917 by the University of Chicago

GOODE'S SERIES OF BASE MAPS
HENRY M. LEPPARD, EDITOR

FIGURE 5.15

The global conveyer belt system shows how deep-water formation, deep currents, and surface currents act together to warm northern Europe's climate.

however, carried by the Trade Winds out of the Atlantic, crosses narrow Central America to fall as rain over the Pacific Ocean, thereby lowering its salinity.

Next, surface waters near Iceland and Greenland, with their relatively high salinities, are chilled and further evaporated by cold dry winds from the Arctic. Both chilling of the waters and addition of salts excluded from sea-ice formation make the surface waters denser. Consequently, they sink below the surface, forming the North Atlantic Deep Water (NADW), which then flows southward along the margin of the western Atlantic basin.

After reaching Antarctica, the North Atlantic Deep Water flows around the continent, carried by deep currents. It is further chilled and receives more salt, excluded by formation of sea ice. In addition, this water mixes with Antarctic water (1 part NADW, 2 parts Antarctic water); thus, by the time the NADW has gone halfway around the continent, it has lost its identity. From there, deep waters (now called Antarctic Bottom Water, AABW) flow northward into all three major ocean basins.

These slow-moving bottom waters gradually warm, therefore becoming less dense, and mix with overlying water masses. Some warming is caused by heat from the ocean bottom and some by interactions of waters with hot, recently erupted rocks on the East Pacific Rise. These waters gradually return to the surface (details of this process are poorly understood), especially through coastal and equatorial upwelling zones.

Surface waters return to the Atlantic, some flowing between Indonesia and Australia, crossing around the Cape of Good Hope (southern tip of Africa) back into the Atlantic. Eventually the surface waters reach the Greenland Sea, where they begin the cycle again. The entire cycle takes about a thousand years.

The concept of the global conveyer belt pulls together many ocean features and their climatic effects. Apparently, the conveyer-type circulation (in its present form) operates at some times and not at others. In other words, the ocean's deep, thermohaline circulation can operate in more than one state. When the conveyer is operating (as it is now), the vigorous thermohaline circulation in the Atlantic warms northern Europe. At other times, the conveyer may not operate because low-salinity surface waters in the Atlantic do not form dense water masses. Under those conditions, no deep waters would form in the North Atlantic, and the subsurface currents in the North Atlantic and the North Pacific would be much more similar than they are at present. Consequently, northern Europe would be much colder than it is now because it would have lost a major heat source. Some speculate that the prolonged cold periods in northern Europe, such as the Little Ice Age (1650–1850), may have been caused by diminished deep-water formation in the Greenland Sea.

QUESTIONS

1. List some major eastern and western boundary currents.
2. Explain why upwelling is common along the equator.
3. Describe the Coriolis effect.
4. What is a geostrophic current? Are all major ocean currents geostrophic?
5. Describe the general deep-ocean circulation patterns.
6. What evidence supports the statement that the oceanic and atmospheric circulations are closely linked?
7. Draw a simple current gyre. Label the eastern and western boundary currents as well as the east-west currents for the North Atlantic and South Pacific oceans.
8. Where is the densest bottom water in the ocean formed? How is it formed?
9. Describe how sound is used to study deep-ocean water masses.
10. Discuss the effect of the global conveyer belt on the climate of northern Europe when it functions and when it cannot function.

SUPPLEMENTARY READINGS

Books

Barry, R. G., and Chorley, R. J. *Atmosphere, Weather, and Climate.* New York: Holt, Rinehart and Winston, 1970. Introduction to climatology.

Stowe, Keith S. *Ocean Science.* New York: John Wiley & Sons, 1979. Elementary, good discussion of physical processes.

Articles

Gordon, A. L., and Comiso, J. C. "Polynyas in the Southern Ocean." *Scientific American* 258(6):90–97.

McDonald, J. E. "The Coriolis Effect." *Scientific American* 186(5):72–78.

Stewart, R. W. "The Atmosphere and the Ocean." *Scientific American* 221(3):76–105.

KEY TERMS AND CONCEPTS

Surface currents
Gyres
Antarctic Circumpolar Current
Equatorial currents
Gulf Stream System
Kuroshio
Western boundary currents
Eastern boundary currents
Intertropical convergence zone (ITCZ)
Coriolis effect
Laminar flow
Turbulent flow
Ekman Spiral
Net water movement
Convergence
Divergence

Ocean topography
Geostrophic currents
Rings
Meanders
Warm core rings
Cold core rings
Upwelling
Sinking
Water masses
Deep-ocean circulation
Thermohaline circulation
Mixing processes
Benthic boundary layer
Acoustic tomography
Global conveyer belt

6

Waves and Tides

Ocean surfaces are rarely still, with waves continually moving across them. The three most important **wave generators** are winds blowing across the ocean surface, earthquakes, and the gravitational attraction of the Sun and Moon. Winds generate most surface waves, ranging from ripples less than 1 cm (0.4 in.) high to giant, storm-generated waves more than 30 m (100 ft) high. Tides also behave like waves but are so large that their wavelike characteristics are easily missed. Seismic sea waves, caused by earthquakes, are still less common. At sea they are easily overlooked but damage coastal communities and cause catastrophic losses of life, especially in lands bordering the Pacific. In this chapter, we examine various types of waves and their behavior in the ocean.

6.1   IDEAL PROGRESSIVE WAVES

We begin by studying a series of waves, called a **wave train.** Progressive waves (in which the wave form moves) passing a fixed point show a regular succession of crests (highest points) and troughs (lowest points). Wave height (H) is the vertical distance from a crest to the next trough (Figure 6.1). Successive crests (or troughs) are separated by one wave length (L). The time required for successive crests (or troughs) to pass the fixed point is the wave period (T), commonly expressed in seconds. Wave periods are often used to classify waves.

Wave speed (V) is calculated by the simple ratio $V = L/T$. In simple words, this formula tells us that wave speed, wave length, and wave period are all directly related; knowing any two factors, we can calculate the third. Wave

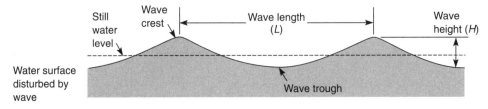

FIGURE 6.1
A simple wave and its parts. Note that the water surface varies above and below its still-water level as the wave passes.

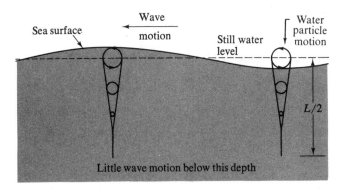

FIGURE 6.2
Wave profile and water-particle motions caused by a wave moving through deep water. Note the diminishing size of the orbits traced by water-particle movements with increasing depth below the water surface.

height (H) is unrelated to the other three factors and must be observed. Wave steepness, expressed as H/L, is the ratio of wave height to wave length.

A cork floating on a water surface moves forward as wave crests pass and backward as troughs passes. After each wave passes, the cork returns to its initial position. This shows that only wave forms move; in deep water there is little net water movement associated with a wave's passage. As we see later, shallow waters are moved by waves breaking on beaches.

Movements of small floats at various depths in a tank show that water parcels below the surface move in nearly circular orbits as waves pass (Figure 6.2). At the surface, orbital diameter equals wave height. Beneath wave crests, particles move in the direction of wave motion; in wave troughs, particle motion is reversed. Below the surface, speed or orbital motion decreases and orbits become smaller. At a depth of half a wave length ($L/2$), wave-induced orbital motions vanish. In **deep-water waves,** wave motions do not affect the bottom.

Water particles move slightly faster in wave crests than in wave troughs, and so there is a slight net movement of water in the direction of wave travel.

Water movement, however, is much slower than wave speed, and for most considerations of wave processes we ignore these small displacements.

6.2   SHALLOW-WATER WAVES

Where water depths are less than one-half wave length ($L/2$), waves interact with the bottom. Water particles near the bottom can move only horizontally, not vertically (Figure 6.3). Farther from the bottom, water particles move in flattened elliptical orbits that become flatter near the bottom and more circular near the surface.

When water depth exceeds one-half wave length, waves are unaffected by the bottom. Therefore, in the deep ocean, a wave's speed is determined by its wave length and period; longer waves move faster than shorter ones. Long waves from distant storms arrive first, followed by shorter waves.

In shallow water, depth controls wave speed. Where the water depth is less than 1/20 of the wave length, wave speed (V) is controlled by average water depth (d) and can be calculated by the formula:

$$V = 3.1\sqrt{d},$$

where V is wave speed (velocity) in meters per second and d is depth in meters.

As waves move from deep water into shallow water, wave speeds and wave lengths change, but not wave periods. Directions of wave advances are also deflected when waves encounter shallow or irregular bottoms; this is called **wave diffraction** (Figure 6.4).

As waves approach beaches and ocean depths decreases (becoming less than $L/2$), orbital water-particle motions are flattened as they interact with the bottom. Although wave period remains unchanged, wave length is shortened. Consequently, wave height increases and wave crests become more peaked. Wave steepness (H/L) increases until it reaches a critical value, about 1/7. When wave crests peak sharply, they become unstable and break (Figure 6.5). Waves usually break when water depth is about 1.3 times wave height.

FIGURE 6.3
Motions of water particles caused by passage of shallow-water waves. Note that the orbits are influenced by the bottom in shallow waters.

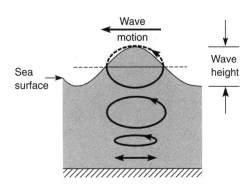

FIGURE 6.4
Waves approaching shores are diffracted by varying water depths, moving more slowly in shallow waters. In deeper waters of bays, wave crests are diffracted and their energy is spread over larger areas. Thus, wave energy is concentrated on headlands, increasing erosion there and depositing materials in bays where there is less wave energy to remove them.

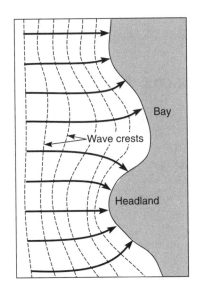

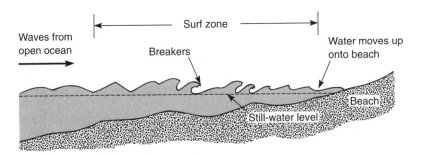

FIGURE 6.5
Waves peak upon entering shallow waters. At a depth of 1.3 times wave height, waves break, re-form, and break again. Finally the water moves up as a thin sheet onto the beach.
[After U.S. Army Waterways Experiment Station, *Shore Protection Manual*]

Breaking waves can form new, smaller waves, which also break as they reach shallower water. Thus, surf zones may have several sets of breakers, depending on wave conditions and bottom configurations.

When waves break, their energy is dissipated, primarily by conversion to heat. If heat from breaking waves were not thoroughly dissipated in large volumes of seawater, surf-zone water temperatures would rise appreciably. Breaking waves also generate waves in the solid Earth, which are detected (as noise) by seismographs.

6.3   SEISMIC SEA WAVES (TSUNAMIS)

Large waves are generated by sudden movements of the ocean bottom, usually caused by earthquakes or explosions. These seismic sea waves, called **tsunamis** (a Japanese word meaning harbor waves; pronounced soo-NA-mhees), have wave lengths up to 200 km (125 mi), periods of 10 to 20 minutes, and wave heights in the open ocean up to 0.5 m (1.5 ft). Although sometimes called tidal waves, they are unrelated to astronomical tides, which we discuss later in this chapter. Tsunamis are examples of free waves; once generated they move independently of the force that caused them.

On the open sea, seismic sea waves are small and pass unnoticed by ships. Sea level rapidly changes, however. When these waves encounter shallow bottom topography they can form enormous breakers. For example, a large valley can focus a seismic sea wave, causing an enormous breaker. Large loss of life and extensive property damage have resulted from tsunamis.

Pacific coastal and island locations are especially vulnerable to these waves. In the past 150 years, the Hawaiian Islands have averaged one seismic sea wave every four years. An international tsunami warning network operates, warning vulnerable locations that a large earthquake, possibly one that might generate a tsunami, has occurred and predicting when a tsunami (if generated) would reach various locations. The network has reduced loss of life, but improvements are needed to reduce the number of false alarms.

New seismographs installed on the sea floor near especially active earthquake zones will improve tsunami detection and predictions. Also, more powerful computers permit more accurate predictions of wave propagation across ocean basins. Hopefully this will reduce the incidence of false alarms experienced under the present system. The problem is, of course, made worse as coastal populations continue to increase.

6.4   WIND WAVES

Gentle winds blowing across a water surface form small wavelets or ripples, less than 1 cm (0.4 in.) high with rounded crests and V-shaped troughs. Because the ripples are so small, surface tension, resulting from the mutual attraction of water molecules, influences their shape. These ripples (also called **capillary waves**) move with the wind and last for only short periods of time, but they provide much of the wind's grip on water surfaces.

As wind speed increases, small gravity waves form from the ripples and travel in the same direction as the winds that formed them. Wave size depends on wind speed, the length of time it blows in one direction, and the distance (called the **fetch**) it has blown across the water (Figure 6.6). In short, the size of wind waves generated depends on the amount of energy imparted by the wind to the water surface.

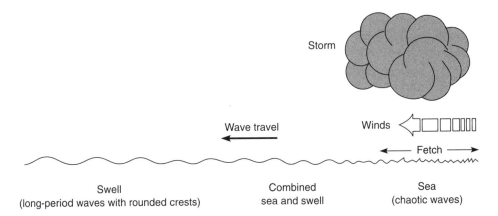

Storm

Wave travel

Winds

Fetch

Swell
(long-period waves with rounded crests)

Combined
sea and swell

Sea
(chaotic waves)

FIGURE 6.6
Winds in a storm form areas of short, choppy, chaotic waves, called seas. As the waves move away from the storm, they develop into waves with longer periods and smooth crests, called swell.

In a storm, a mixture of chaotic waves and ripples, known as a **sea,** develops. These waves continue to grow until they are as large as wind of that speed can generate. After the winds die, the waves move away from the generating area and become more regular. Long, regular waves are known as **swell** (Figure 6.6).

Wind waves are classified according to their periods. Ripples have periods of a fraction of a second. Wind waves in fully developed seas have periods up to 15 seconds; swells have periods of 5 to 16 seconds. Unlike currents, wind waves are not affected by the Coriolis effect, because little water is moved by passing waves.

Waves move initially in the same direction as the winds that caused them. Winds from different directions can destroy earlier sets of waves and generate new ones. Little energy is lost by waves as they cross over the deep ocean. Thus, waves continue until they meet an obstacle where their energy is dissipated on a cliff, beach, or breakwater. Waves generated in Antarctic storms have been detected near the Aleutians off Alaska, nearly halfway around Earth.

6.5 STATIONARY WAVES

In contrast to progressive waves, which move across water surfaces, stationary waves (in which wave forms do not move) occur widely in the ocean; they are also called **seiches** (pronounced say-shees; French for sloshes). They are easily generated in a full, round-bottomed coffee cup by tilting the cup and setting it down on a surface. Viewed from the side, the water surface tilts toward one side and then toward the other. This oscillation of water surfaces is caused by the

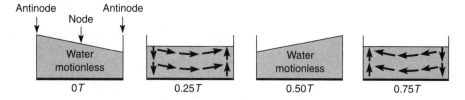

FIGURE 6.7
Water motions in a simple stationary wave, shown at quarter-period intervals. The cycle
shown here repeats itself.

standing wave you generated. When you spill coffee from a cup you are carrying, the culprit is often a standing wave.

During each oscillation, the water surface remains at the same level at certain spots called **nodes** (Figure 6.7). Stationary waves in a small cup or dish
usually have only one nodal line where the water level does not change. But it
is also possible to have several nodal lines or nodal points about which the water
surface tilts.

At the **antinodes,** or crests, vertical water movements are largest. Many
antinodes or crests can occur in a container or body of water, but they always
occur at the ends. By placing chips or dye in the water, we can observe water
motions generated by stationary waves. There is no water movement when the
water surface is tilted most. When the water surface is horizontal (its equilibrium position), the water is moving most rapidly. The largest horizontal water
movements occur below nodal lines; beneath the crests, wave movements are
entirely vertical.

Stationary waves are generated in enclosed water bodies by sudden distur-
bances, such as storms or sudden changes in atmospheric pressure. Once set in
motion, a water body will oscillate with its period determined by water depth
and basin length. Eventually, seiches die out owing to loss of energy through
friction of the waters moving along the basin's edges. Lake Erie has a characteristic seiche with a period of 14.3 hours; that is, every 14.3 hours the lake surface
returns to its original position. In Lake Michigan, the characteristic seiche has a
period of 6 hours. Because of Earth's rotation, wave crests in large lakes in the
Northern Hemisphere move clockwise around the basins.

6.6   INTERNAL WAVES

Internal waves (Figure 6.8) occur at interfaces between layers of different densities. Some are easily detected by the presence of slicks (streaks of calm water);
these are convergences caused by internal waves where debris and surface-
active substances such as oils and greases collect.

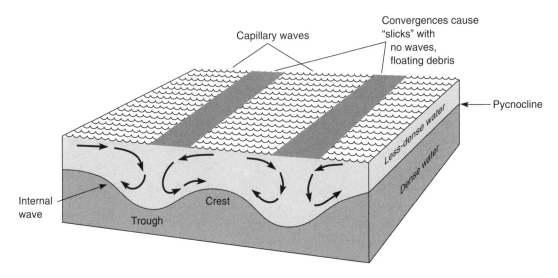

FIGURE 6.8
Internal waves cause surface-water movements that collect floatable materials and sur-
face-active substances (oils, soaplike compounds) above the troughs. These materials
inhibit formation of capillary waves, causing slicks (areas of ripple-free waters) that mark
the troughs of the internal waves.

Temperature or salinity measurements in the pycnocline often reveal internal waves, because internal waves cause periodic shallowing or deepening of
waters having characteristic temperatures or salinities. Tidal phenomena cause
some internal waves in the ocean. Because of the difficulty of observing internal
waves in the deep ocean, they are poorly understood.

6.7   ENERGY FROM WAVES

Ocean waves are a potential energy source. A single wave 1.8 m (6 ft) high in
water 9 m (30 ft) deep has about 10 kW of power in each meter of wave front.
The power dissipated by waves breaking on a beach during a single storm is
enormous. Power generated by waves has been used for decades to power whis-
tles and gongs for navigational buoys. The problem is to develop ways of extracting large amounts of energy at reasonable cost and transmitting or storing them
for future use.

6.8   TIDES

Among ocean phenomena, tides (periodic rising and falling of sea surfaces) are
easy to observe. A firmly anchored pole with marked heights can be used to

measure relative heights of the sea surface at frequent intervals, and the rise and fall of the tide can be determined from such data. A simple tide gauge consists of a float connected to a pencil, which draws a tidal curve (a record of sea level over several days) on a paper-covered, clock-driven cylinder. Modern tide gauges measure water pressure at the bottom (this indicates water depths) and record the data for later analysis, usually by computers. Some gauges transmit tidal readings by satellite to a central facility, where the data are analyzed immediately to detect tsunamis.

Although partially understood since antiquity, the astronomic origin of the tides was first explained in detail by Sir Isaac Newton's (1642–1727) law of gravitational attraction, which states that the attraction between two bodies is directly proportional to the product of their masses and inversely proportional to the square of the distance between them. In other words, the attraction between bodies increases as the masses of the bodies increase, and decreases as the distance between them increases. The effect of this observation is that it makes possible tidal predictions based on movements of the Sun and Moon and local modifications of these tides by complicated shorelines. In short, tides behave much like clocks, and Newton provided the theory to permit tidal predictions.

To understand tides, we must consider the gravitational effects of the Sun and Moon. These two are most important to ocean tides because of the Sun's large mass and the Moon's nearness to Earth. Newton's theory of gravity can be used to develop an **equilibrium tidal model** for an idealized, water-covered Earth without continents. First we consider tidal effects caused by the Moon alone.

Tides are caused primarily by gravitational attractions among Earth, Sun, and Moon. Tides occur in the solid Earth, in ocean waters, and in the atmosphere, but we will consider only ocean tides. Ocean tides are caused by slight differences along Earth's surface in the gravitational attraction and centrifugal forces between Earth and Moon. (We will ignore the Sun for the moment.) The ocean's surface is deformed by these forces into an egg-shaped envelope. Earth rotates within its deformed water envelope (Figure 6.9). An observer would experience this as the rise and fall of the tides at the observer's location.

On the side of Earth nearest the Moon, ocean water is drawn toward the Moon because the distance between the two is slightly less than at Earth's center. As a result, the water surface is pulled by gravity, forming a bulge on the side nearest the Moon. On the side opposite the Moon, the gravitational attraction is slightly less than at Earth's center. These centrifugal forces deform the water surface, forming a bulge on the side opposite the Moon. The two tidal bulges are areas of high tide, and between them are troughs or areas of low tide (Figure 6.9).

The Moon passes over any location once every 24 hours, 50 minutes (one lunar or tidal day). On a water-covered Earth, any point passes beneath two tidal crests and two tidal troughs during each tidal day. If the Moon remained in the plane of Earth's equator, the two high waters at each location would be equal. However, the Moon's position (and associated tidal bulges) shifts 28.5° north of the equator to 28.5° south of the equator. This changes the relative

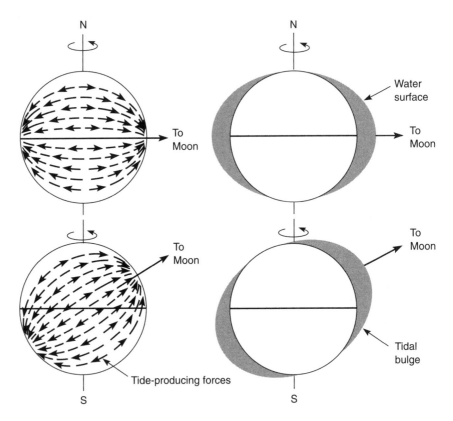

FIGURE 6.9
Tide-producing forces and resulting tidal bulges on a water-covered Earth with the Moon in the plane of Earth's equator (upper) and above the plane of the equator (lower).

heights of high and low waters at any point. When the Moon is not in the plane of the equator, there will be one high tide and one low tide each day.

The Sun also affects tides, but to a lesser degree than the Moon does. Interactions between the effects of the Sun and Moon account for some of the complexity involved in predicting tides. At certain times during the Moon's travel around Earth, the Sun and Moon act together (Figure 6.10). At these times, the bulges, or crests, are highest, and water levels in the tidal troughs between them are lowest. These are called **spring tides,** when daily tidal ranges (vertical distance between high and low tides) are largest. When the Moon is near its first and third quarters, solar and lunar tides partially cancel each other, and daily tidal range is lowest. These tides are called **neap tides.**

Newton's theory explains the relative effects of the Sun and the Moon on ocean tides. It also explains why there are two high and two low waters each day in many locations. But it fails to predict two important aspects of the tides:

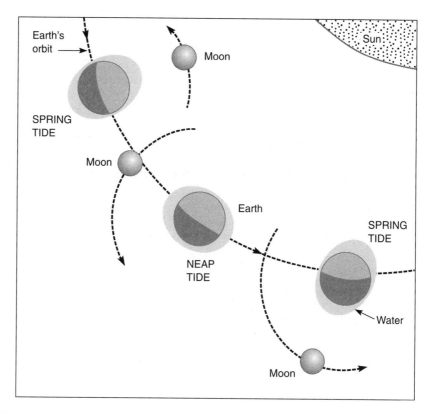

FIGURE 6.10

Relationships of Earth, Moon, and Sun during spring and neap tides. Note that the Moon and Sun are on a line with Earth during spring tides but act at right angles to each other during neap tides.

their height, and the times of high and low tides relative to the Moon's passage overhead.

Consider the timing of high water. Our simple equilibrium model predicts that high tides will occur when the Moon is highest in the sky above our position or directly below our position on the other side of Earth. This requires that each tidal bulge travel about 1650 km/hr (1000 mi/hr) to keep pace with the Moon.

Tides are very long waves. The two tidal bulges, or crests, occur on opposite sides of Earth, and so tides have wave lengths of nearly 22,000 km (14,000 mi). The ocean is slightly less than 4 km (2.5 mi) deep; thus, tides behave like shallow-water waves. Tidal waves could keep up with the Moon only if the ocean were at least 22 km (14 mi) deep. Consequently, tidal crests are displaced from their equilibrium positions by frictional drag on the ocean bottom and by Earth's rotation (Figure 6.11).

FIGURE 6.11
The position of the tidal bulge is determined by the equilibrium between the Moon's gravitational attraction and the frictional drag of a rotating Earth (viewed from the North Pole).

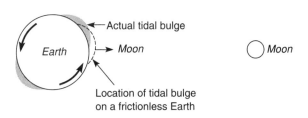

6.9 OCEAN RESPONSE TO TIDE-GENERATING FORCES

If Earth were smooth and water-covered, and ocean basins had simple shapes, tides could easily be predicted, following Newton's laws. As we know, however, Earth is only partly water-covered and is far from smooth, and ocean basins have a variety of sizes and shapes and bottom topographies. Therefore, the water in each basin responds differently to tide-generating forces.

Tidal height and timing vary greatly from basin to basin and even at different places within the same basin. Tides in ocean basins have been compared to a set of stringed instruments designed by a mad artist. The strings on all the instruments are tuned to the tide-generating forces, but the different sizes and shapes of the instruments produce a great variety of sounds.

Tides in ocean basins respond to a few rules:

1. If the characteristic period of the standing wave in a basin is short relative to the period of the tide-generating forces, there is ample time for water levels to be displaced in step with the tide-generating forces. Such a basin has an equilibrium tide.
2. If the characteristic period of the standing wave is very long relative to the period of the tide-generating forces, there is not enough time for water levels to keep step with the tide-generating forces. In this case, the tides are small and reversed. In other words, low tide occurs when we would have predicted high tide (and vice versa), based on equilibrium tide theory.
3. When the characteristic period of a standing wave in a basin is nearly the same as the tide-generating forces, high and low tides occur nearly as we would have predicted, but tidal heights are much greater than predicted. The closer the correspondence between the two, the larger the tidal range.

Tides at any locality are combinations of standing and progressive waves. In some basins, one or the other predominates. In long, narrow basins such as the Red Sea and Long Island Sound, the tide behaves like a standing wave with nearly simultaneous rise or fall from one end of the basin to the other; high or low tide occurs everywhere at nearly the same time. Timing of tidal phenomena is controlled by the tide entering as a wave from the adjacent open ocean.

In Chesapeake Bay or Puget Sound, the tide behaves like a progressive wave. High water occurs first at the entrance and then advances inland like the crest of a progressive wave. This takes many hours, and there may be several crests (or high-water areas) in the system at any time. Each crest (high water) is separated by low water, corresponding to the trough of a simple wave.

A similar situation occurs in each ocean basin. Tides move through the ocean like progressive waves. Timing of the tides is controlled by considerations that we have discussed, but paths of tidal waves and their behavior in the basin are strongly affected by local effects.

Tidal prediction requires a long period of observations (many months to many years) at a particular point. From these observations, it is possible to determine how that ocean area responds to the tide-generating forces and how this response is affected by complexities of basin shapes, by waves coming from adjacent ocean basins, by friction of waters moving in the basin, and by Earth's rotation. Tidal tables, prepared years ahead, are perhaps the most successful predictions of oceanic phenomena.

6.10   TYPES OF TIDES

There are three types of tides: diurnal, semidiurnal, and mixed. **Diurnal tides** (one high water and one low water per tidal day) are simplest (Figure 6.12); they are common in the northern Gulf of Mexico and in the Pacific near Southeast Asia. **Semidiurnal tides** (two high waters and two low waters per tidal day) are common on the Atlantic coasts of North America and Europe (Figure 6.13). Note that successive high-water and low-water levels are approximately equal, as predicted by Newton's equilibrium theory. Along the Pacific coast of North America, **mixed tides** are most common. In mixed tides, successive high-water and low-water stands differ appreciably, as shown in Figure 6.14. Thus, there are higher high water and lower high water, as well as higher low water and lower low water. The complexity of long-term tidal phenomena can be seen in tidal curves for one month at New York in Figure 6.15 (semidiurnal tide) and at Seattle in Figure 6.16 (mixed tide).

FIGURE 6.12
Diurnal tides have only one high tide and one low tide per day. Such tides are rare in North America, occurring primarily in the Gulf of Mexico.

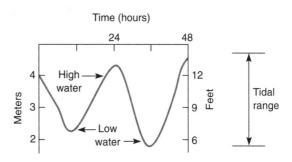

FIGURE 6.13
Semidiurnal tides have two roughly equal high and low tides per day. They are common along the Atlantic coast of the United States.

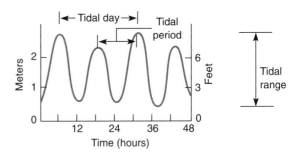

FIGURE 6.14
Mixed tides at (a) Seattle and (b) Honolulu. Note how the tidal currents (F = flood current, E = ebb current) are associated with the different stages of the tide. Mixed tides are common on the Pacific coasts of the Americas.

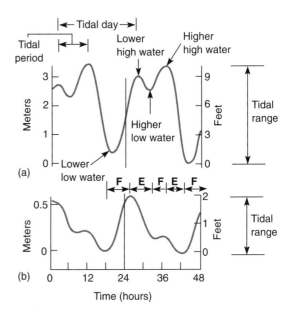

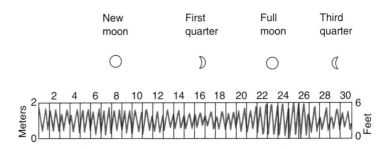

FIGURE 6.15
Variations in tidal heights during one month of semidiurnal tides at New York. Note the relationships of the spring and neap tides to the phases of the Moon.

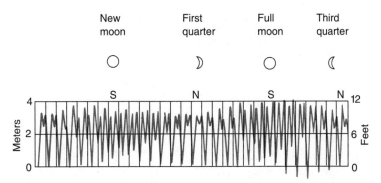

FIGURE 6.16
Variations in tidal heights during one month of mixed tides at Seattle.

6.11 TIDAL CURRENTS

Like other waves, tides cause horizontal water movements, known as **tidal currents.** Away from obstructions, tidal currents constantly change direction (Figure 6.17) and are thus called rotary currents. Rotary tidal currents repeat this cycle once each tidal period.

In nearshore areas and in rivers and harbors, coastlines obstruct tidal currents (Figure 6.18), preventing rotary tidal currents. Here we find reversing tidal currents (Figure 6.19): currents flow in one direction during part of the tidal cycle and then reverse their direction of flow during the remainder of the cycle.

When water levels rise in harbors, water flows toward the land (called **flood tides**). As tidal currents flow seaward, sea levels fall; these are called **ebb tides.** Periods of **slack water** (little or no currents) separate ebb and flood currents (Figure 6.19).

FIGURE 6.17
In rotary tidal currents, current directions and speeds continually change during one tidal period (12 hours) at Nantucket Shoals off the New England coast. Times shown are in hours before and after low and high tide.

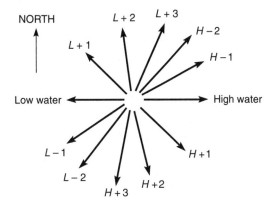

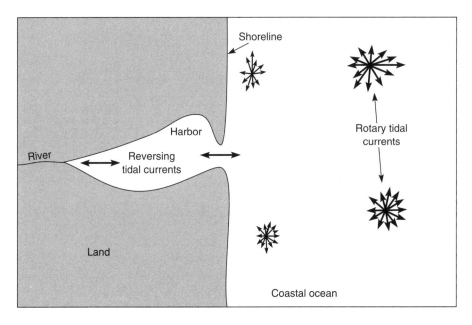

FIGURE 6.18
Relationships between offshore rotary tidal currents and reversing tidal currents in bays
and rivers. Note that near the coast, tidal currents are transitional between rotary and
reversing tidal currents. Also note that the strongest currents parallel the coast.

FIGURE 6.19
Reversing tidal currents at
Admiralty Inlet, Puget Sound,
Washington. Note that flood
currents generally correspond
to rising (incoming) tides,
whereas ebb currents occur
during falling (outgoing) tides.

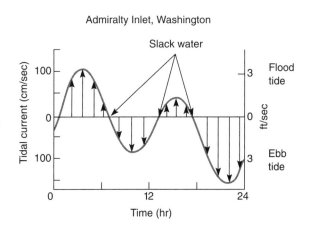

Predictions of tidal currents are based largely on experience, rather than on
theory. Tidal currents are strongly affected by local and regional winds and river
flows, which can overwhelm astronomical forces. In addition, the time and
speed of maximum flood or ebb current can vary substantially within a single
bay or harbor. Tidal current tables contain predictions of these currents, usually

based on long series of current measurements, and assuming that astronomical forces predominate. Thus the predictions are often unsatisfactory.

Many ports operate their own tide gauges and current meters so they can advise ships about conditions as they change. The difference of a fraction of a meter in water depth or in current strength and direction can mean the difference between a ship entering or exiting port safely and going aground. This is especially critical as port traffic increases and ships become larger.

6.12   ENERGY FROM TIDES

For centuries, humans have harnessed tides to produce power. In 1650, for instance, Boston had a tidally powered mill for grinding corn. Most modern tidal-power units generate electricity. In the early 1990s, three large, electrical, tidal-power installations were operating: one on the Rance Estuary in Brittany (northwestern France), another on Kislaya Bay (northern Russia near Murmansk), and the third at Annapolis Royal on the Bay of Fundy (Atlantic Canada). Their high construction costs and problems of power transmission or storage limit the number of tidal-power stations.

Tidal-power plants use changing sea level during tidal cycles to create a head of water (elevated surface) during high tide. Water then runs through the turbines at low tides, generating electricity. Thus, in planning tidal-power plants, efforts are made to locate the greatest tidal ranges, exceeding 5 m or 16 ft (Figure 6.20), to obtain useful amounts of energy. On land, dams are built on rivers to provide heads of about 100 m (300 ft) or more. There are no areas having such large tidal ranges; the largest is 16 m (50 ft) on the Bay of Fundy, in eastern Canada.

Most tidal-power plants (Figure 6.21) involve one or more dams closing off a bay from the ocean. The larger and wider the opening to be dammed, the more expensive its construction. Most potential tidal-power sites are in higher latitudes, where glaciers have cut deep, narrow embayments and scoured landscapes down to bedrock. But these sites are rare and are often prized for other purposes, such as for their scenic beauty or resources.

Furthermore, there are problems with the times of maximum power generation, because tidal-power generation is tied to tidal cycles, which shift 50 minutes later each day. Availability of peak tidal power rarely comes at times of peak demand for electricity. Several schemes have been developed to get around this problem. One is to connect tidal-power plants to form a large power distribution network so that the power generated can be used somewhere regardless of the time at which it is generated. Another proposed solution is to build several dams and to store water at high tidal levels so that one basin is a reservoir and another is a collector. This proposal is expensive and calls for large, often scarce sites. Still another scheme is to store power for later use when it is needed. One way to store energy at times of low demand is by pumping water to high reser-

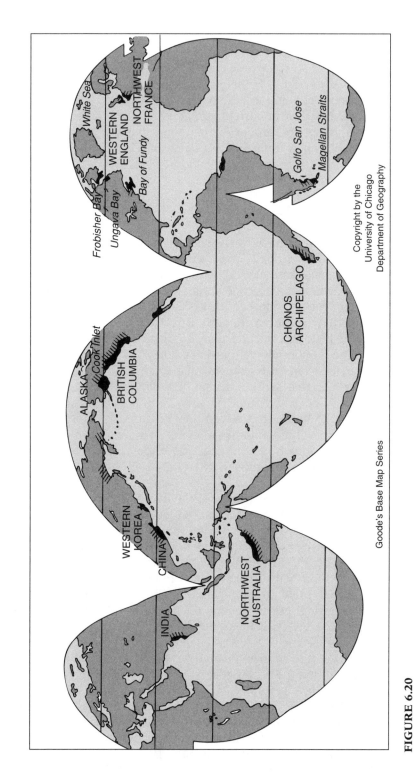

FIGURE 6.20

Potential sites for tidal-power generating stations, where average spring-tidal ranges exceed 5 m (16 ft).

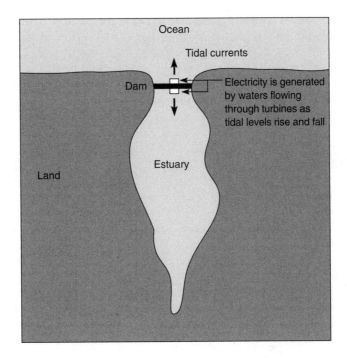

FIGURE 6.21
A tidal-power plant requires a dam to enclose an inlet or estuary. Tidal currents flow through the turbines to generate power. Such sites are scarce and often used for other purposes.

voirs (so that when power is needed, the water can be released to generate power). Another possibility is making synthetic fuel, such as hydrogen, and storing it until it is needed or transporting it to some other place.

6.13   STORM SURGES

Storm surges are sea-level changes caused by strong winds, which are usually associated with major storms such as hurricanes or typhoons. Around open-ocean islands, storms cause only small sea-level changes (usually less than a meter) because, with no boundaries, water flows are basically unrestricted. Storm surges are essentially the opposite of upwelling in that waters are brought to coastlines by winds and accumulate there until the winds die down and the waters associated with the storm surge flow away.

On low-lying coasts, storm surges can raise sea level many meters, causing catastrophic flooding. Such surges killed 250,000 people in 1864 and 1876 (in what is now India and Bangladesh) around the Bay of Bengal in the northern Indian Ocean. In 1953, a severe North Sea storm surge raised sea level about 4 m (13 ft), flooding both low-lying Dutch and British coastal areas. The high water levels of the storm surge combined with high waves overtopped and breached many dikes protecting the Dutch coast. In the Netherlands, about 25,000 km² (9600 mi²) were flooded, 2000 people were killed, and 600,000 had to be evacuated. After this disastrous storm, barriers were constructed on both the Dutch and British coasts to protect against future storm surges. The costs of constructing these barriers ran into many billions of dollars.

QUESTIONS

1. Why are tsunamis most common in the Pacific?
2. What three processes generate waves in the ocean?
3. During an eclipse of the Moon, would we have spring or neap tides? Why?
4. List and briefly discuss the two major types of wind waves. Indicate the dominant force that determines a wind wave's shape and speed.
5. Contrast an ideal stationary wave with an ideal progressive wave.
6. Explain how the equilibrium theory of the tide explains semidiurnal tides.
7. List some of the wave characteristics that distinguish seas from swells.
8. Compare a storm surge and wind-induced sinking along a coast.
9. Why are suitable tidal-power sites so rare?
10. Explain why tidal-power stations are expensive to build.
11. What can be done to protect low-lying coastal areas against flooding caused by storm surges?

SUPPLEMENTARY READINGS

Books

Bascom, W. *Waves and Beaches: The Dynamics of the Ocean Surface,* rev. ed. Garden City, NY: Anchor Books, Doubleday and Company, Inc., 1980. Elementary.

Clancy, E. P. *The Tides: Pulse of the Earth.* Garden City, NY: Doubleday, 1969. Elementary.

Russell, R. C. H., and MacMillan, D. H. *Waves and Tides.* New York: Philosophical Library, Inc., 1953.

Tricker, R. A. R. *Bores, Breakers, Waves and Wakes.* New York: American Elsevier Publishing Company, Inc., 1965.

Articles

Bascom, W. "Ocean Waves." *Scientific American* 201 (2):89–97.

Goldreich, P. "Tides and the Earth-moon System." *Scientific American* 226(4):42–57.

Truby, J. D. "Krakatoa—The Killer Wave." *Sea Frontiers* 17(3):130–139.

KEY TERMS AND CONCEPTS

Wave generators
Progressive waves
Wave trains
Deep-water waves
Shallow-water waves
Wave diffraction
Tsunamis
Wind waves
Capillary waves
Fetch
Sea
Swell

Stationary waves
Seiches
Nodes
Antinodes
Internal waves
Equilibrium tidal model
Types of tides: diurnal, semidiurnal, mixed
Tidal currents: flood tides, ebb tides
Slack water
Energy from tides
Storm surges

7

Life in the Ocean

Marine organisms obtain the energy and materials they need to build tissue from their surroundings—a complex of interrelated systems called a marine **ecosystem**. Life began in the ocean, and most of the many forms of life that ever existed on Earth lived, or still live, in the ocean. Thus, marine life is both ancient and extremely diverse. This diversity is important when we search for new organisms and new compounds to meet human needs.

In this chapter, we examine various aspects of oceanic life, selected marine organisms, the factors that control their distributions and abundance, and some effects of marine organisms on the ocean's abundance of nutrients (substances necessary for growth) and the composition of seawater.

7.1 LIVING CONDITIONS IN THE OCEAN

We begin by comparing conditions for oceanic life (Figure 7.1) with those for life on land. On land, organisms live primarily on or near the surface; only birds, insects, and the largest trees extend more than a few meters into the atmosphere. In contrast, the ocean is a vast, three-dimensional environment for life (Figure 7.1). About 99.5% of the inhabited living space on Earth is in the ocean; only about 0.5% is on land. Thus, living space in the ocean greatly exceeds that on land and also provides many different kinds of habitats unfamiliar to us as land dwellers. We know most about the relatively shallow parts of the ocean, and least about the mid-depths.

Most marine plants and animals live near the ocean's edges—its surface, bottom, and coastal areas. Most food is produced in the ocean's sunlit, near-sur-

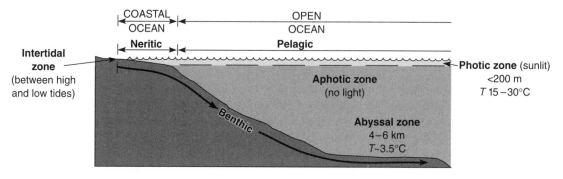

FIGURE 7.1
Major life zones in the ocean. Availability of light and food as well as temperature ranges primarily control where organisms live in the ocean.

face waters, called the **photic zone,** and thus organisms of all kinds are most abundant there. After animals feed, they void the remains, forming rapidly sinking **fecal pellets** that transport food to the ocean depths and eventually to the bottom.

Organisms living in near-surface waters also sink to the bottom after death. Other animals and bacteria consume these dead organisms as they sink or as they lie on the ocean floor. Thus, organic matter produced in surface waters sustains life throughout the ocean and feeds bottom-dwelling organisms; only a small fraction of the organic matter produced near the surface is finally buried in sediment deposits on the bottom.

Seawater and living tissues are nearly identical in density—about 1.03 g/cm^3 (0.6 oz/in.^{3}). This provides two advantages to marine organisms. First, they can float (or sink slowly) without expending much energy. Consequently, many marine organisms are planktonic (floating) or free swimming. Second, marine organisms do not need the structures land plants and animals require to support themselves. Because they do not need such support mechanisms, many marine organisms are quite delicate and jellylike (composed largely of water), unlike anything living on land. Jellyfish are familiar and beautiful examples.

Ocean water is nearly opaque to light, and so visibility is greatly limited. Even with bright lights and exceptionally clear waters, humans can rarely see more than 30 m (100 ft) in the ocean, and usually less. In contrast, we commonly see for many kilometers through clear air. Marine animals that depend on sight to locate and capture food must have large, sensitive eyes. Smelling and sensing vibrations from other organisms are widely used instead by marine animals to locate food. As we know (Chapter 3), seawater is nearly transparent to sound, and so many animals detect their prey using sound, much as submarines detect other submarines. Indeed, some whales are able to communicate with each other across entire ocean basins.

Many organisms emit light. These organisms have special organs, usually involving bacteria, that produce light through chemical reactions. This is called **bioluminescence.** Some use light to attract and eat unwary organisms. Others use light to frighten predators. For instance, squids, when alarmed, release clouds of luminous ink to confuse predators and thus help them escape. Other organisms apparently use their light organs to attract mates.

Marine organisms range in size from tiny cells too small to be seen by light microscopes (Figure 7.2) up to the largest animals on Earth, the blue whale, which grows to be 30 m (100 ft) long. Most marine organisms are small, averaging about the size of a mosquito, and are mostly moved by currents, because they are too small to swim effectively against them. Some marine animals have mechanisms for catching these smaller drifting organisms, such as structures made of mucus. Others, called filter feeders, have special structures that capture food by filtering large volumes of water.

The diversity of oceanic life varies substantially from place to place. Marine organisms are most diverse in tropical waters, where there are many different types of organisms but few representatives of each group. Conversely, in severe and changeable conditions (subpolar regions, for instance) there are relatively few different kinds of organisms but many individuals of the types that can tolerate the harsh conditions.

Organisms are distributed unevenly in the ocean. Large expanses of open-ocean waters, far from land, are nearly devoid of life; these areas have excep-

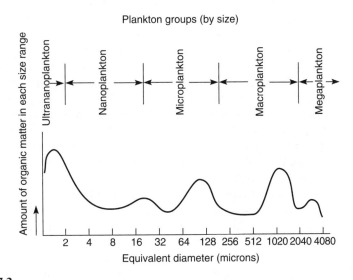

FIGURE 7.2
Relative abundances of different sizes of planktonic marine organisms.
[After T. R. Parsons, M. Takahashi, and B. Hargrave, *Biological Oceanographic Processes*, 2d ed. (Oxford: Pergamon Press Ltd., 1977). Used by permission]

tionally clear surface waters, appearing a luminous blue color in the sunlight. The Sargasso, (western side of the North Atlantic subtropical gyre), the Mediterranean Sea, and large areas of the open ocean have such unproductive but colorful waters.

Coastal ocean waters over continental shelves, called **neritic environments,** are especially rich in marine life. These relatively shallow waters produce more plant material than do comparable open-ocean areas. In addition, plant materials wash into the ocean from land. Thus, about half of the world's fish production is caught in the coastal oceans, mostly in upwelling areas.

Marine organisms usually occur in patches. Some **patchiness** is caused by processes that concentrate floating organisms into long rows, much the same way that tidal currents sweep floating debris into long stripes, called tide rips. In other cases, patchiness results from plants growing fastest where nutrients are most abundant. An abundance of phytoplankton usually attracts animals to feed, increasing the patchiness.

7.2   MARINE ORGANISMS

Marine organisms are classified as **plankton, benthos, nekton,** and **neuston,** depending on where and how they live. Each has different lifestyles and plays different roles in the ocean, as we shall see.

Oceanic plankton include minute, mostly microscopic, pelagic plants and animals that are easily moved by ocean currents, going where the water takes them. Marine plankton are further divided into phytoplankton (plant plankton) and zooplankton (animal plankton). The size ranges and abundances of the major groupings of planktonic organisms are shown in Figure 7.2.

Phytoplankton are microscopic, single-celled plants. Two major groups, diatoms and dinoflagellates (Figures 7.3 and 7.4), produce most of the organic matter in the ocean, which supports marine animals. Requiring sunlight for photosynthesis, phytoplankton must live in sunlit, near-surface waters, known as the photic zone. The thickness of the photic zone is controlled by latitude, season, time of day, and water clarity. The dark interior of the ocean is known as the **aphotic zone.**

Zooplankton (Figures 7.5 and 7.6) includes two groups: holoplankton, which are animals that spend their entire lives as plankton, and meroplankton, which are planktonic only during certain, usually larval, developmental stages. Larval stages of fishes are also among the meroplankton. Adult forms of meroplanktonic larvae usually live in or on the ocean bottom. In shallow, continental shelf waters, meroplankton are more numerous than in deep, open-ocean waters. Currents disperse the planktonic larvae of benthic organisms, allowing them to colonize new areas when conditions are favorable. Zooplankton include representatives of most major groups of marine animals, but are dominated by

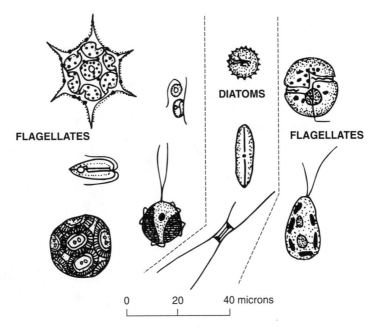

FIGURE 7.3
Common marine plankton. Flagellates and diatoms produce much of the food that supports the smallest organisms in the ocean.
[After T. R. Parsons, M. Takahashi, and B. Hargrave, *Biological Oceanographic Processes,* 2d ed. (Oxford: Pergamon Press Ltd., 1977). Used by permission]

small crustaceans, including shrimplike copepods, amphipods, mysids, and euphausiids (Figure 7.6).

Zooplankton can live at all depths in the ocean. They are most abundant in surface and near-surface waters, where food is most plentiful. Oceanic zooplankton are classified according to the water depths they inhabit:

Epipelagic: near-surface waters, less than 200 m (660 ft) deep
Mesopelagic: mid-depths, between 200 and 700 m (660 and 2300 ft) deep
Bathypelagic: deep waters, more than 700 m (2300 ft) deep

Deeper-water zooplankton are usually larger than near-surface organisms.

Many mesopelagic organisms migrate vertically every day. They feed in near-surface waters at night, returning to darker, deeper waters during daylight hours; this is called **diurnal migration.** Presumably they migrate daily to avoid predators that need light to locate their prey.

Benthic organisms live on or in the ocean bottom (benthic environment). Some organisms (**infauna**) live in sediments near the bottom. Others (**epifauna**)

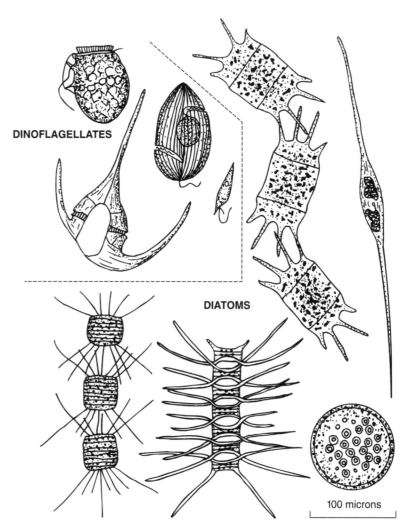

DINOFLAGELLATES

DIATOMS

100 microns

FIGURE 7.4
Common marine phytoplankton: dinoflagellates and diatoms. These organisms produce much of the food that supports marine ecosystems.
[After T. R. Parsons, M. Takahashi, and B. Hargrave, *Biological Oceanographic Processes,* 2d ed. (Oxford: Pergamon Press Ltd., 1977). Used by permission]

live on the bottom or in near-bottom waters. Some organisms burrow; others attach themselves to the substrate; and still others crawl about on the bottom or swim freely in near-bottom waters. Members of nearly every major group of marine animals are represented in the zoobenthos (Figure 7.7), among them protozoans, various worms, mollusks, crustaceans, and starfish. Bacteria are abundant in sediments and also live free in ocean waters, although most bacteria live their lives attached to particle surfaces. Bacteria decompose much of the organic matter in the ocean, returning its constituents in dissolved form back into the water. (There will be more about this when we discuss nutrient cycles.)

In relatively shallow waters, plants grow on the bottom. Sea grasses and marine algae are most abundant in shallow waters. None grow below about 200 m (660 ft), because they need sunlight for photosynthesis. (We discuss kelp, the largest marine plant, in the next chapter.)

Benthic organisms occur more abundantly in shallow waters than on the deep-ocean floor, because of the greater abundance of food there. In near-shore waters, **standing crops** (**biomasses**) of benthic organisms may comprise as much living matter as 1 kg/m^2 (0.2 lb/ft^2). For comparison, in the deep ocean, living matter on the bottom amounts to no more than 1 g/m^2 (0.03 oz/yd^2).

In **anoxic** bottom waters (lacking dissolved oxygen) in isolated ocean basins (such as the Black Sea), normal marine animals cannot survive; only bacteria and a few sulfide-tolerant organisms can live there. Similar situations occur

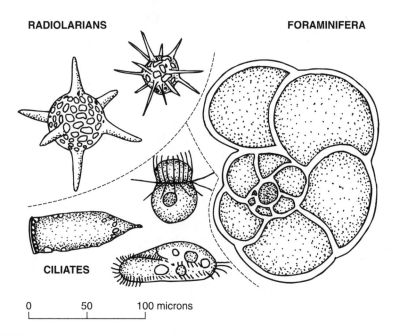

FIGURE 7.5
Marine zooplankton: radiolarians, ciliates, and foraminifera. Shells of these relatively large organisms are major contributors to sediment deposits.
[After T. R. Parsons, M. Takahashi, and B. Hargrave, *Biological Oceanographic Processes*, 2d ed. (Oxford: Pergamon Press Ltd., 1977). Used by permission]

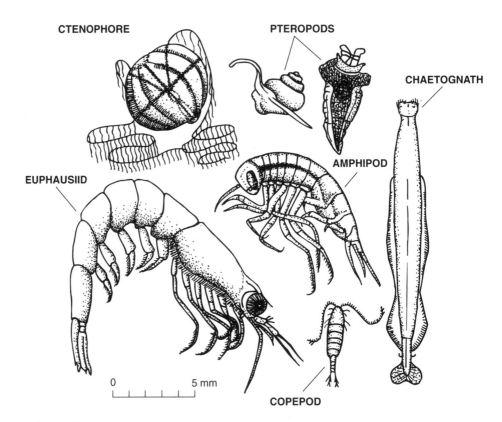

FIGURE 7.6
Larger zooplankton organisms include both herbivores and carnivores. These organisms are well known, because they are involved in the food webs that support major fisheries.
[After T. R. Parsons, M. Takahashi, and B. Hargrave, *Biological Oceanographic Processes,* 2d ed. (Oxford: Pergamon Press Ltd., 1977). Used by permission]

where pollution or excessive plant growth (called **eutrophication**) has depleted dissolved oxygen from near-bottom waters in harbors and estuaries and some coastal waters.

    Nekton are animals that are strong enough to swim independently of currents. Nekton (Figure 7.8) include marine mammals (whales and porpoises), most fish, cephalopod mollusks (especially squids), and some swimming crustaceans (crabs, for instance). Nekton are most abundant in near-surface waters but live at all depths in the ocean. Like some mesopelagic organisms, many nektonic organisms migrate daily from the aphotic zone into near-surface waters to feed at night.

    Neuston are animals living at the sea surface, depending on water's surface tension for support. Jellyfish and related forms, such as *Physalia* (Portuguese Man-of-War) and *Velella* (the By-the-Wind-Sailor), as well as less common floating snails, called pteropods (Figure 7.9), are members of the neuston.

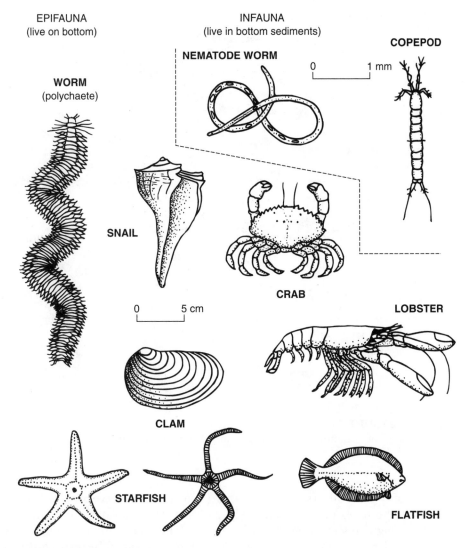

FIGURE 7.7
Common benthic (bottom-dwelling) organisms include representatives from many different groups.

7.3   MARINE ECOSYSTEMS AND FOOD WEBS

Marine ecosystems consist of producers (plants), consumers (animals), and decomposers (bacteria, fungi). In ecosystems, nutrients and energy from the Sun are made into food by photosynthesizing organisms. The food they produce is consumed by other organisms, which eventually die and decompose, releasing nutrients and starting the cycles again (Figure 7.10).

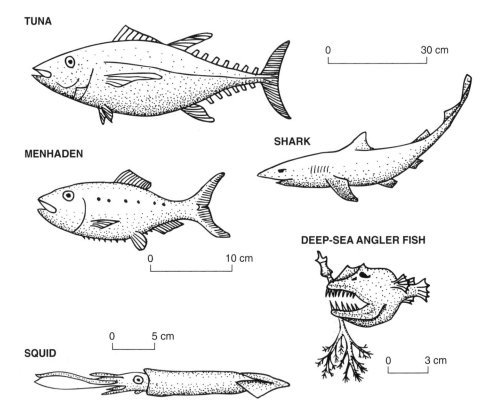

FIGURE 7.8
Some common nektonic organisms. Many of these organisms are exploited by various fisheries.

Producers, usually plants (also called **autotrophs,** "self-nourishing"), are the bases of ecosystems (Figure 7.11). Using solar energy, their photosynthetic pigments form energy-rich carbohydrates. Consumers (animals), which depend directly or indirectly on plants for food, are called **heterotrophs.** Animals feeding directly on plants are called **herbivores;** those that prey on other animals are called **carnivores. Omnivores** eat both plants and animals. After death, organisms are decomposed by **detritivores,** usually bacteria and fungi, which decompose cells and return nutrients back to the water.

Feeding relationships, called **food chains,** can be quite simple. In a hypothetical case on land, grass (primary producer) is eaten by a cow (herbivore), which in turn is eaten by a tiger (carnivore) or a human (omnivore).

A simple oceanic food chain is:

Diatoms >>> Copepods >>> Herring >>> Humans
1000 g   100 grams   10 g   1 g

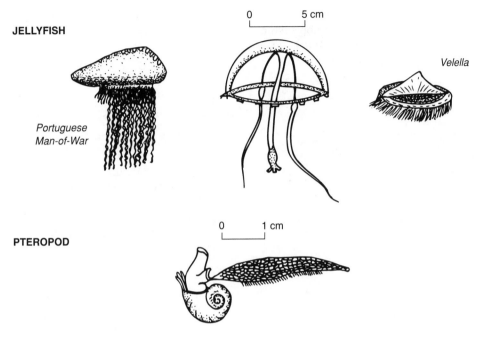

FIGURE 7.9
Some common neustonic (surface-dwelling) organisms. Jellyfish are seasonally abundant and often wash up on beaches.

In a food chain, each stage is called a **trophic level.** Only about 10% of the energy is transferred from one level to the next higher one. Thus, 1000 g of diatoms is required to produce 10 g of herrings, which in turn produces 1 g consumed by humans.

Simple food chains occur in some lakes but rarely in the ocean. Organisms usually eat many different kinds of food, and in turn are eaten by a host of others. These complicated feeding relationships are called **food webs.** A relatively simple one is shown in Figure 7.12.

7.4   LIGHT

Energy from the Sun supports plant growth in near-surface ocean waters. As we learned in Chapter 4, surface waters absorb most incoming solar radiation, changing its energy to heat. Only about 0.1% of the incoming radiant energy is used by plants, but this energy is sufficient to produce the food for virtually all life in the ocean. Most plant growth occurs at depths where light levels exceed 1% of incoming solar radiation at the surface (Figure 7.13). Thus, virtually all plant growth occurs in the photic zone, near the ocean surface.

FIGURE 7.10
Patterns of phytoplankton and zooplankton abundance in various open-ocean provinces.

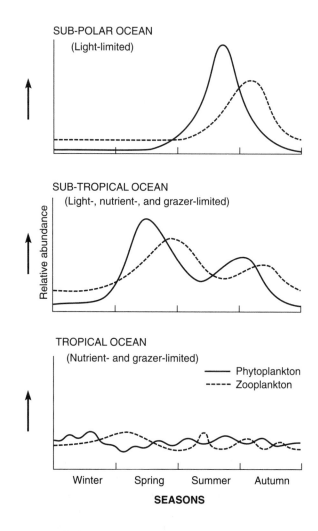

SUB-POLAR OCEAN
(Light-limited)

SUB-TROPICAL OCEAN
(Light-, nutrient-, and grazer-limited)

Relative abundance

TROPICAL OCEAN
(Nutrient- and grazer-limited)

—— Phytoplankton
----- Zooplankton

Winter   Spring   Summer   Autumn

SEASONS

Even in the clearest seawaters, light penetrates only the surface zone. At depths around 100 m (about 300 ft), nearly all visible light has been absorbed or scattered. In subtropical regions in relatively particle-free waters with no dissolved organic substances, less than 0.01% of the insolation remains as visible light at about 200 m (about 700 ft), usually blue-green (Figure 7.13).

A striking feature of many open-ocean areas is the intense, almost luminous blue color of the waters, indicating that they are particle free. These are oceanic "deserts," where surface waters cannot support abundant phytoplankton growth. (We discuss the reasons for this in the next section.) The water's blue color is caused by light scattering and absorption. Pure water scatters bluish colors more readily than reddish colors. Water is also more transparent to blue light than to red. Thus, water is blue for the same reason that the clear sky is blue, because of light scattering.

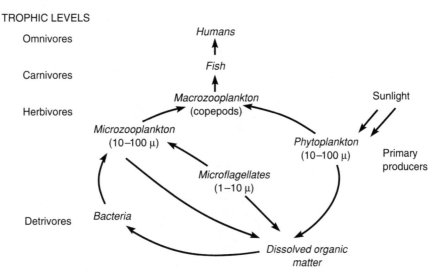

FIGURE 7.11
Simplified food web for oceanic life. Note that there are many interactions involving only the smallest forms. This is sometimes called the microbial loop. Only a minute part of the food produced in the ocean goes into fishes and other organisms consumed by humans.

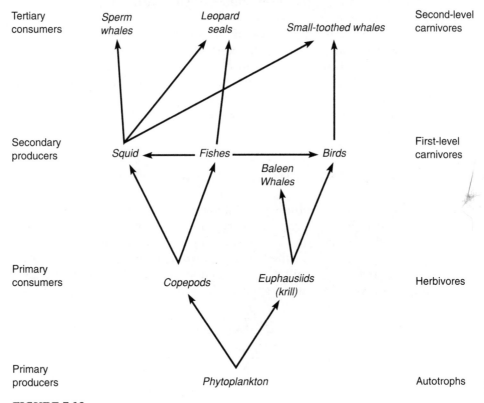

FIGURE 7.12
Simplified food web in Antarctic waters. Note that a single group of organisms, such as squid or krill, may be eaten by several kinds of animals.

FIGURE 7.13
Amounts of different colors
of light remaining at various
depths in the open ocean,
after incoming solar radiation
is mostly absorbed in near-
surface waters. Note that the
scale of light remaining
changes by factors of ten.

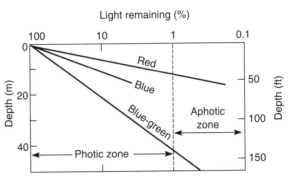

Particles and dissolved materials can discolor seawaters. Coastal waters are often brownish, greenish, or even reddish, depending on the particles or organisms involved. For instance, red-colored organisms (such as dinoflagellates) cause **red tides,** waters discolored by enormous numbers of planktonic organisms. Certain red tide organisms release various toxic substances that can poison both marine animals and humans who eat them. This is an increasingly severe problem in many coastal areas.

Suspended particles and dissolved materials also limit light penetration. In turbid seawater at depths of 10 m (33 ft), light levels may be comparable to those at 100 m (330 ft) in clear seawater, and the remaining light is yellow-green. In many tropical and subtropical wetlands, decomposing plant materials release dissolved substances that color waters brownish or blackish, and further limit light penetration. These materials are often carried by currents into coastal oceans.

7.5   PHOTOSYNTHESIS AND RESPIRATION

Plants photosynthesize, combining carbon (from carbon dioxide in seawater) and water, by means of energy from sunlight, to form energy-rich compounds such as carbohydrates (sugars and starches). Phytoplankton produce most of the organic matter in the ocean. **Photosynthesis** and **respiration** can be represented by the following, reversible chemical equation (in other words, the process can go either way):

PHOTOSYNTHESIS (forms carbohydrates)
>>>>>>>>>>>>>>

$6CO_2 + 6H_2O + energy <<—>> C_6H_{12}O_6 + 6O_2$
Carbon dioxide   Water   Carbohydrate   Oxygen gas

<<<<<<<<<<<<<<<
RESPIRATION (uses carbohydrates for energy)

To obtain energy from food, animals essentially "burn" carbohydrates, a process called respiration, using oxygen and releasing carbon dioxide and water while obtaining energy—the reverse of photosynthesis.

At this point, it is worth noting another major contrast between productivity on land and in the ocean. If you look at a pasture on land, you see lots of grass and only a few cows grazing on it. In other words, there is far more grass (in mass) than cows.

In the ocean, this situation is reversed. If you look into clear ocean waters, you will usually see a few fish but no evidence of the phytoplankton that supply their food. Indeed, the mass of herbivores (grazers) can exceed the mass of phytoplankton present.

The reason is simple. Phytoplankton produce more, cell for cell, than plants on land. Diatoms grow, reproduce, and die much faster than do grass and trees, their terrestrial equivalents. Thus, oceanic herbivores can consume huge quantities of phytoplankton without exhausting the supply. If marine algae were only as efficient as their land-based equivalents, you would always be looking at green waters and you would rarely be able to see fish or other animals.

We discuss the role of grazers in controlling phytoplankton production, along with limited sunlight and nutrients, in the next section.

7.6   NUTRIENTS AND PRODUCTIVITY

In addition to light, phytoplankton must extract many dissolved substances from seawater in order to grow and reproduce. Some substances, such as inorganic carbon (from dissolved carbon dioxide), calcium, sodium, potassium, magnesium, and sulfate, are abundant everywhere in ocean waters. Others, known as **nutrients** (specifically nitrogen and phosphorous compounds, and silica), are scarce in seawater. In surface waters, where sunlight is available, nutrients are removed from the water by plant growth.

After phytoplankton die and decompose, the nutrients in their tissues are returned to the water. About 90% of the phytoplankton decompose and release their nutrients in surface waters, thereby sustaining further growth by phytoplankton there. Decomposition takes time; the nutrient releases are not instantaneous.

About 10% of the cells sink below the pycnocline before decomposing. Their nutrients are thus unavailable to sustain immediate further growth in the photic zone. (Remember that the pycnocline inhibits upward water movements.) Consequently, about 10% of the nutrients are released in subsurface aphotic waters where there is too little light for photosynthesis. These nutrients accumulate there, enriching nutrient concentrations below the pycnocline. This is called the ocean's **biological pump** (Figure 7.14), which we also discussed in Chapter 3.

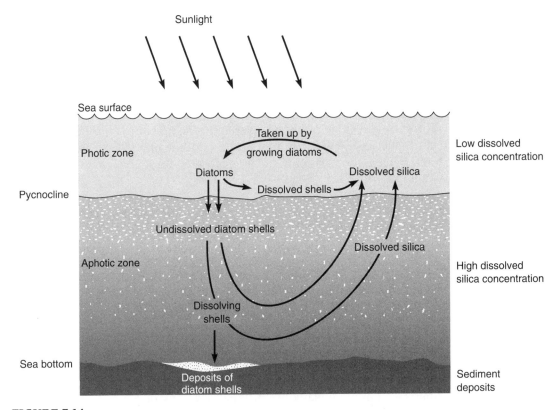

FIGURE 7.14
The biological pump functions by removal of materials (silica in this example) from near-surface waters by photosynthesizing organisms. About 10% of the plankton sinks below the pycnocline before decomposing. This enriches deep-ocean waters in these substances, which are then returned to the surface by upwelling and other processes. The same process also works for various nutrients and carbon dioxide, which are removed from near-surface waters and enrich the deeper ocean waters.

Deep waters, rich in nutrients, return to the photic zone by means of vertical water movements, usually upwelling or mixing. Near-surface productivity over most of the ocean is controlled by the rates at which nutrient-rich, subsurface waters return to supply organisms growing in the photic zone.

Coastal ocean areas are especially productive (Figure 7.15). There, deep, subsurface waters and their nutrients are mixed back into the photic zone by storm winds, waves, and currents flowing over irregular ocean bottoms.

Estuarine circulation (discussed further in Chapter 8) also brings nutrients to surface waters. Estuaries also retain particles, planktonic organisms, and nutrients. Such food-rich conditions are favorable for the growth of larvae of many organisms and account for the highly productive commercial and recreational fisheries in estuaries.

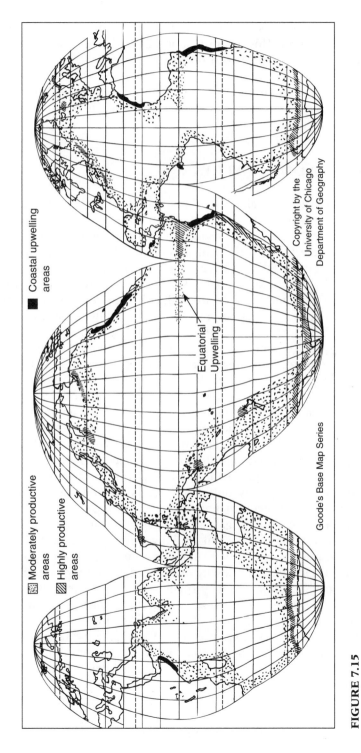

FIGURE 7.15

Distribution of productivity in the ocean. Note the highly productive coastal and equatorial upwelling areas.

These return flows of subsurface waters to the surface support prolific phytoplankton growth. This, in turn, supports productive fisheries where small fish (sardines and anchovies) feed directly on large phytoplankton (diatoms) and zooplankton.

Along the equator, nutrients return to the surface through upwelling. Thus, equatorial upwelling areas are also highly productive of phytoplankton and fishes (Figure 7.15).

7.7   SEASONAL VARIATIONS

Seasonal changes in availability of light and nutrients control the abundances of phytoplankton and zooplankton (Figure 7.10), especially plankton blooms (unusually high concentrations of cells). During winters in the middle and high latitudes, there is too little light to support abundant phytoplankton growth. Beginning in early spring, increased insolation warms surface waters, producing a less dense (and therefore highly stable) surface layer. This helps retain phytoplankton in the photic zone by inhibiting their mixing down into aphotic waters. Because there is little photosynthesis during winter (too little light), nutrients are available in surface waters to support plant growth.

The resulting rapid increase in phytoplankton abundance, called a **phytoplankton bloom,** in turn supports rapid zooplankton growth, eventually leading to a **zooplankton bloom.** Because zooplankton do not reproduce or grow as fast as phytoplankton, there is also a time lag between a phytoplankton bloom and a subsequent zooplankton bloom. Eventually, the abundant zooplankton eat phytoplankton, reducing their abundance and often ending the phytoplankton bloom. As zooplankton food becomes more scarce, the zooplankton bloom also ends. (A rapid decrease in abundance is called a population crash.) As light intensity drops in late autumn, phytoplankton growth and reproduction decrease, and eventually both phytoplankton and zooplankton abundances diminish to the low levels that are typical of winter.

Rapid phytoplankton growth also depletes nutrients in near-surface waters down to levels where they can no longer reproduce. Later, the abundance of zooplankton drops as the food supply is reduced by grazing. Sometimes storms mix near-surface waters, bringing more nutrients into the photic zone. If there is still enough light, other, usually smaller, blooms occur before plankton growth drops to winter levels in response to light limitations. This situation is most pronounced in coastal waters and in the North Atlantic.

In tropical waters, there is enough light all year round to sustain photosynthesis. Thus, nutrient supplies control phytoplankton abundances. Storms mix the waters, causing small blooms, but none as intense as the spring blooms that occur in higher-latitude waters. Here the continual abundance of zooplankton continually graze phytoplankton populations, which cannot rapidly increase to form massive blooms. Small blooms can occur because phytoplankton grow much faster than zooplankton. In most parts of the ocean, phytoplankton abundances are controlled by zooplankton grazing so that dramatic blooms of either phytoplankton or zooplankton are rare.

Another factor, only recently discovered, is that diseases of planktonic organisms can also limit blooms. Viruses occur in abundance in seawater. Under certain circumstances, they can infect and destroy both phytoplankton and zooplankton. Indeed, virus infections have been implicated in causing extinctions of certain phytoplankton that occur in dense concentrations, such as coccoliths.

7.8   MARINE MAMMALS

Marine mammals (whales, dolphins, seals, and walruses) are warm blooded, breathe air, bear their young live, and later nourish them with milk produced in mammary glands. All are legless and have streamlined bodies and horizontal tail flukes. They feed on other marine organisms and spend most, if not all, of their lives in the ocean. Because of their intelligence and ability to learn, marine mammals often appear in marine-life shows and are popular in aquaria.

There are two kinds of whales. Baleen (whalebone) whales are filter feeders (Figure 7.16), have no teeth, and swim slowly—normally 3 to 5 km/hr (2 to 3 mi/hr). They feed by filtering planktonic crustaceans or small fishes from the water. They take a mouthful of water and expel it through parallel plates of hornlike baleen, and the small animals are filtered out as the water is expelled through the baleen. Then, using their tongues, they scrape the food off the baleen and swallow it. The largest whales (including blue, right, humpback, and gray whales) are all filter-feeding, baleen whales. Blue whales, the largest animals that have ever lived (up to 150 tons), eat up to three tons per day of small shrimplike crustaceans, called krill. (The largest sharks also feed on plankton and small fish in the same way.)

Toothed whales are smaller, swim faster (about 20 km/hr, or 12 mi/hr), and feed on fish and squid. To locate their prey, toothed whales use echo loca-

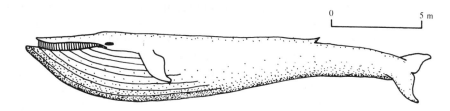

FIGURE 7.16
Blue whales are the largest animals on Earth. They grow to be 30 m (100 ft) long and weigh 150 tons. These baleen whales filter plankton and small fishes from the water. Only a few hundred are thought to have survived uncontrolled whaling. It is not clear if the species can recover from such low numbers of individuals.

tion systems (much as sonar systems locate submarines). They have specialized organs to emit and detect the sounds. Humans are still learning from whales how to use sound in the ocean.

Because of their various foods, marine mammals occupy different positions within a single food web. In a simplified Antarctic food web (Figure 7.12), for instance, whales and seals include both first-level carnivores (baleen whales) and second-level carnivores (leopard seals, toothed whales, and sperm whales).

Some large whales feed in subpolar waters on the large summer zooplankton blooms and the benthic organisms they support. In fall, these whales migrate to warmer subtropical waters to give birth to their young. Their gestation period is about one year. Young whales gain weight rapidly in preparation for the long trip back to their feeding grounds. Gray whales are regularly seen migrating south along the U.S. Pacific coast between November and March. They apparently return northbound farther offshore and are thus less visible from land.

Toothed whales have gestation periods of more than a year, and baleen whales about 11 months. Whales usually give birth to one calf at a time. A fin whale calf is about 6 to 7 m (20 to 23 ft) long at birth, about one-third the mother's length. It suckles for about 6 months and grows to about 13 m (about 40 ft). In 3 to 8 years, fin whales reach sexual maturity at about 20 m (65 ft). Whales live from 40 to 100 years. With such long life spans, it takes many decades to rebuild whale stocks once they have been reduced to low levels.

Hunting of whales for oil and meat began as far back as the ninth century, and was important in Europe during the Middle Ages. Whaling later became important in America during the seventeenth, eighteenth, and nineteenth centuries. The U.S. whaling industry reached its peak in 1850, with more than 700 active American vessels. Whale oil was used for lamps until it was replaced by kerosene, early in the twentieth century. Baleen was used for corset stays, among other purposes.

These uncontrolled whale harvests reduced some coastal whale species (right whale, California gray whale) to near extinction. We still are not sure if enough animals survived to rebuild all the stocks, even though commercial whaling stopped in the late 1980s in most countries. A few species, such as the California gray whale, have apparently recovered. Limited whaling is permitted by aboriginal peoples, such as those in Arctic coastal villages, where whaling is an important part of their culture and a major food source.

Using acoustic listening arrays, scientists can now learn about distributions and abundances of whales throughout the ocean. It may soon be possible to make better counts of whales and to assess the status of whale stocks.

7.9   SEABIRDS

Seabirds (Figure 7.17) are conspicuous and familiar forms of marine life. Some spend nearly their entire lives at sea; others can barely walk on land. Penguins, for example, leave the ocean only to lay eggs, and to hatch and rear their young.

FIGURE 7.17
Brown pelicans are fish-eating birds that feed in nearshore waters.
(Courtesy National Oceanic and Atmospheric Administration)

Some seabirds, however, live primarily on land, using the sea only for food. For example, long-billed curlews probe exposed beach sands to eat tiny, deeply buried animals. Similarly, short-billed shorebirds eat smaller epifauna that live closer to the beach surface. Such birds are significant predators; an oyster catcher can eat as many as 300 clams per day.

Some seabirds actively pursue their prey beneath the water. Penguins and cormorants actively swim underwater. Pelicans, on the other hand, fly over the water; when they sight fish, they plunge from several meters to snatch it from the water surface.

Because seabirds must drink seawater, they have special glands to secrete excess salt through their nostrils. Diving birds have folds of skin to keep water out of their nostrils while diving. To keep their feathers dry, most seabirds have oil-secreting glands near the base of the tail. To waterproof their feathers, they spread oil over them with their beaks. However, a few birds, such as cormorants, lack oil-secreting glands and must spread their wings to dry after diving. Still others, such as frigate birds, catch their food on the wing, snatching it from just below the surface to avoid getting wet.

Seabirds locate food by sight, smell, and sound. They depend on finding concentrations of food, such as schools of fish. Many seabirds live in large colonies near coastal upwelling areas on islands near the equator, on coasts, or near ice margins; all of these are areas where plankton and fish productivity is high. Birds also follow ships, eating garbage or animals brought to the surface in the ship's wake.

Fish-eating seabirds, such as pelicans (Figure 7.17) are especially vulnerable to chemical pollutants. To supply the energy needed for flying, they must eat large amounts of fish, and thus they ingest whatever toxins or poisons the fish have accumulated in their fat tissues. Because the birds are long-lived (pelicans live 25 years or more), they accumulate these toxins in their tissues over many years.

Brown pelicans on both the Pacific and Atlantic coasts were decimated in the 1960s and 1970s by the pesticide DDT, which caused thin eggshells. Few young were hatched because adult birds crushed their eggs while incubating them. Consequently, pelican populations decreased drastically, both in numbers and in distribution ranges. Now that the use of DDT has been banned in the United States, these pelican populations are recovering and returning to their original ranges. A similar situation has occurred around the Great Lakes among bald eagles, which are land-dwelling but fish-eating birds.

7.10   DEEP-OCEAN-BOTTOM ORGANISMS

The types and abundances of abyssal benthic organisms are controlled by the availability both of food and of suitable space on the bottom, whether it is rocky or covered by soft sediments. Nearly all bottom-dwelling organisms depend on food produced by photosynthesis in near-surface waters, even though they may be thousands of meters below the photic zone. Only a few percent of the food produced in the photic zone reaches the bottom.

Most food reaching the bottom comes in rapidly sinking fecal pellets voided by zooplankton in surface waters. Rarely, large dead animals, such as whales, also reach the bottom, where they support dense communities of benthic organisms for years. Thus, benthic animals must be able to locate and utilize these scarce food supplies. Details of how they do this are still unclear. Deep-sea animals are generally small, are slow-growing, and reproduce infrequently, usually after one of their rare meals.

7.11   HYDROTHERMAL-VENT COMMUNITIES

Hydrothermal vents on recently erupted volcanic rocks provide special environments that are exploited by highly specialized animals (Figure 7.18). First, the recently erupted rock provides hard bottoms needed by some animals for attachment, as well as an abundant supply of nutrients and energy that allows food production in the absence of sunlight. Therefore, active vents are surrounded by large populations of specialized organisms—oases of life on a normally sparsely populated ocean bottom.

The food for these organisms comes not from photosynthesis but from specialized bacteria that get the energy they need to make food from metals and certain gases (hydrogen sulfide, methane, and others) in vent waters. These bacteria use chemical processes different from those used by photosynthetic plants. The bacterial process is called **chemosynthesis.** Thus, in hydrothermal-vent

FIGURE 7.18
Giant gutless worms grow profusely around an active vent on the mid-ocean ridge near the Galapagos Islands, off South America.
(Photograph by Kathleen Crane, Courtesy Woods Hole Oceanographic Institution)

ecosystems, the primary producers are the specialized bacteria, and they form the bases of the food webs there.

Large (up to 3 m, or 10 ft, long), fast-growing worms (Figure 7.18) having no digestive tracts are conspicuous members of vent communities. These organisms obtain food from bacteria that live inside special organs in their bodies. The worms' tentacles are bright red because their blood uses hemoglobin (like human blood) to transport oxygen to the bacteria; they also have special blood proteins to transport sulfides to the bacteria. Oxygen-breathing organisms lacking such protective systems would be killed by hydrogen sulfide in the vent waters.

Other unusual organisms living around vents are large, fast-growing clams and mussels. Crabs and fish live near the vents, feeding on attached organisms. Few vent systems have been thoroughly explored, and so we are still learning about the various organisms that are capable of living in such environments.

Volcanic activity on the ocean bottom, like everywhere else on Earth, is sporadic, with eruptions separated by long periods of inactivity. Each eruption apparently supports active vents and their associated organisms for years or even decades. When vents no longer discharge fluids, nearby animals starve because without the gases and metals that supply energy, their bacteria cannot produce food. Thus, vent organisms must have larvae that are dispersed widely in order to find other active vents. Larvae of some vent animals are apparently temporarily planktonic before returning to attach themselves at suitable locations on the bottom. Others are transported by near-bottom currents.

Similar organisms grow around cold seeps discharging sulfides or methane-bearing waters on continental slopes. These waters flow through the continental rocks and return to the ocean in slow flows, unlike the spectacular "black smokers." Comparable communities also live near municipal sewage discharges on continental shelves off Los Angeles and San Diego. Again, the chemosynthetic bacteria are able to use materials discharged in the sewage to produce food for their host organisms.

Similar organisms have also been seen living on bones of dead whales lying on the bottom. Apparently as the whales' bodies decompose, they release compounds that these organisms can also utilize. As scientists' ability to explore the deep-ocean floor increases, they are likely to find far more of these communities and many different organisms populating them.

7.12   FISHERIES

Beginning in the 1950s, increased investments worldwide in new vessels and instruments and increased exploration for new fishing grounds caused substantial increases in fish catches (Figure 7.19). Marine fish catches increased about 6% per year until the 1970s. Indeed, many scientists predicted that food from the ocean would feed starving peoples everywhere. Beginning in the 1980s, the

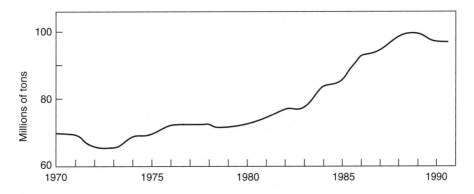

FIGURE 7.19
Increased numbers of fishing vessels and better fish-finding equipment caused production of the world's fisheries to increase until the late 1980s. Since then it has decreased slightly. (Data from *The Economist,* 19 March 1994)

increases continued but were smaller, only 2.3% annually. In 1990, world fish catches decreased for the first time and the future of marine fisheries began to look bleak.

An early indication that the ocean might be not be able to produce unlimited amounts of fish came from collapses of the California sardine fishery in the 1950s and the Peruvian anchovy fishery (the anchovy is a close relative of the sardine) in the early 1970s. At the time it collapsed, the Peruvian fishery was the world's largest. During the 1972 El Niño, the fishery collapsed, falling from about 12 million tons to 1 to 3 million tons per year in the late 1980s. The fishery recovered slightly, but never returned to anything near its peak production levels.

As fishing efforts continued to increase worldwide, many fish stocks were overexploited. To pay for their new ships and equipment, fishermen had to go farther and work longer, while catching smaller amounts of fish. Furthermore, there was a worldwide shift in the types of fishes caught. For example, high-valued but slow-growing fishes, such as Atlantic cod, were so depleted that the Canadian government closed the cod fishery off Newfoundland in 1992. Since then, cod stocks have continued to decline for unknown reasons, and immediate prospects for their recovery are not good. Most fish stocks suffer from overexploitation, resulting from overinvestment and improvements in ships and fish-catching gear, combined with ineffective protection of the stocks.

In tropical waters, similar shifts have also occurred. As catches of top predators, such as groupers and snappers, declined, fishing fleets turned to smaller, short-lived fishes. The only increases in large, high-value fishes came from opening of new tuna-fishing grounds in the western central Pacific and Indian oceans. Species such as squid entered new markets in large quantities, but here, too, production increased primarily because new areas were being exploited.

Much of the fish caught in the industrial fisheries is used to make fish meal for feeding chickens and cattle, rather than being eaten directly by humans. As we saw earlier, using fish protein in this way is wasteful because of the inevitable losses that occur at each stage of the process.

Furthermore, many ocean areas are increasingly polluted. The semi-isolated seas, such as the Mediterranean and Black seas, are polluted by nutrients from agricultural runoff and sewage discharges in their coastal waters. Here, too, fish production has declined. Large lakes and totally enclosed seas such as the Caspian are also suffering severe deterioration.

From these results, it seems certain that the ocean cannot produce unlimited amounts of food. Indeed, the data suggest that fisheries management has failed everywhere to protect the resource—even in the North Atlantic, where we know most about the ocean and the fishes involved. Further, this record of repeated failure casts doubt on whether sustainable development of any of the ocean's living resources can be achieved with our limited understanding and present political structures.

7.13   SUSTAINABILITY

An attractive concept called **sustainability** is aimed at maintaining resources, human welfare, and our natural environment so that future generations will not be penalized by today's activities. As the history of fisheries development shows, we are a long way from sustainability in this and probably all human activities. Our present way of life has been achieved by using resources of all types beyond their capacity to regenerate. And in the process, many resources are being damaged and possibly destroyed. Whaling is a prime example in the ocean; clear cutting of forests is an example on land.

Aquaculture, which is the growing of marine plants and animals for human use, may be one way to use the ocean in a sustainable way. Aquaculture is already a multibillion-dollar industry, primarily in Asia, where freshwater fishes and seaweeds dominate. Marine fishes (salmon) and molluscs (mussels, oysters, clams) are developing sectors of aquaculture.

Unlike in agriculture, which draws on thousands of years of plant and animal domestication, there are no domesticated marine plants or animals that can be grown totally in captivity. Most marine aquaculture depends on capturing wild organisms in their early stages and growing them under controlled conditions. Another problem is our primitive knowledge about the diseases that affect marine organisms, both in the wild and under cultivation. All these facets are being worked on. Even new organisms are being developed through marine **biotechnology,** in which existing plants and animals are modified to make them more suitable for cultivation.

QUESTIONS

1. Define ecosystem. Describe the major functions that organisms perform in ecosystems.
2. Discuss why upwelling conditions can support major fisheries.
3. List the types of marine organisms that are important constituents of phytoplankton.
4. Why are deeper ocean waters richer in nutrients than surface waters after a period of phytoplankton growth?
5. Why can many marine organisms exist without skeletons?
6. Discuss the factors controlling the abundance and distribution of organisms in the surface ocean.
7. Describe a typical grazing food chain.
8. What is the role of encrusting algae in building coral reefs?
9. What substances are produced commercially from marine organisms?
10. Discuss the evidence that the ocean's capacity to produce fish is limited.
11. Why is aquaculture likely to provide more seafood in the future?

SUPPLEMENTARY READINGS

Books

Burton, R. *The Life and Death of Whales*. 2d ed. London: Deutsch, 1980.

George, D., and George, J. *Marine Life: An Illustrated Encyclopedia of Invertebrates in the Sea*. New York: Wiley-Interscience, 1979.

Lalli, C. M., and Parson, T. R. *Biological Oceanography: An Introduction*. Oxford: Pergamon Press, 1993. Modern view of oceanic life.

Lockley, R. M. *Ocean Wanderers: The Migratory Sea Birds of the World*. Harrisburg, PA: Stackpole Books, 1974. Elementary.

Marshall, N. B. *Ocean Life in Color*. New York: Macmillan, 1971. Elementary; illustrated in color.

Parsons, T. R.; Takahashi, M.; and Hargrave, B. *Biological Oceanographic Processes*. 3d ed. Oxford: Pergamon Press, 1984. Advanced reference.

Raymont, J. E. G. *Plankton and Productivity in the Oceans*. 2d ed. Vol. 1, *Phytoplankton*. Oxford: Pergamon Press, 1980. Standard reference.

Sumich, J. L. *Biology of Marine Life*. Dubuque, IA: William C. Brown Group, 1976. Elementary.

Article

Isaacs, J. D. "The Nature of Oceanic Life." *Scientific American* 221(3):146–165.

KEY TERMS AND CONCEPTS

Ecosystems
Major life zones
Photic zone
Bioluminescence
Neritic environments
Patchiness
Plankton
Benthos
Nekton

Neuston
Phytoplankton
Aphotic zone
Zooplankton
Infauna
Epifauna
Bacteria
Standing crops (biomasses)
Anoxic bottom waters

Eutrophication
Autotrophs
Carbohydrates
Heterotrophs
Herbivores
Carnivores
Omnivores
Food chains
Trophic level
Food webs
Red tides
Photosynthesis
Respiration
Nutrients
Decomposition

Biological pump
Blooms
Marine mammals
Baleen whales
Migrations
Toothed whales
Seabirds
Hydrothermal-vent communities
Chemosynthesis
Fishery collapses
Pollution
Fisheries management
Sustainability
Aquaculture
Biotechnology

8

Coasts and Coastal Oceans

Coastal oceans and adjacent seas account for only a small part of Earth's surface (12.5%) and of the ocean's volume (4%). Humans depend on coastal oceans for transportation, recreation, and food. These highly productive waters support about one-half of the world's fish catches, but they also receive large waste discharges, which coastal currents disperse. Thus, coastal oceans—the most important oceanic region to humans—are also most threatened by human activities, such as pollution, massive alterations, overfishing, and general mismanagement. In this chapter we examine coastal-ocean processes, how they differ from those in the open ocean, and how they affect our lives. Then we consider some specific coastal-ocean environments, and the increasing demands made on them, and finally how these affect both coasts and coastal oceans.

8.1   COASTAL OCEANS

Coastal oceans are shallow ocean regions lying over continental shelves. They are strongly affected by nearby lands, river outflows, large human populations, and industrial and agricultural discharges. Coastal oceans are also highly variable. Their currents, water characteristics, and even marine life change over relatively short distances (a few kilometers to a few hundred kilometers) and short periods of time (a few days to a few weeks). In one familiar example, trash discarded in coastal waters by recreational boaters will likely float onto nearby beaches within a few hours. In contrast, in open-ocean regions, processes take years to centuries. For instance, a float from a fishing net carried by surface cur-

FIGURE 8.1

The U.S. Atlantic coast north of Chesapeake Bay was formed primarily by terrestrial processes. South of the Chesapeake, it was formed primarily by marine processes. Glaciers cut Long Island Sound and the New England coast, north of Cape Cod. Delaware Bay and Chesapeake Bay are ancient, flooded river valleys.

(Courtesy NASA)

rents takes several years to cross the North Pacific Ocean. As we have learned, deep-ocean waters take centuries before they return to the surface.

Coastal-ocean waters respond within a few hours to winds blowing over them. Water temperatures and river discharges change over a few weeks or months. Coastal-ocean distances (Figures 8.1 and 8.2) are also shorter than those involving the open ocean.

Many coastal waters are partially isolated from the open ocean. For instance, parts of the Southern California Bight (off Los Angeles) are partially

FIGURE 8.2
In southern California, mountain building has formed rocky shorelines with narrow beaches. The continental shelf is as mountainous as the land, with many silled basins, some filled with sediment deposits. The Channel Islands partially isolate the northern end of the Southern California Bight from the nearby Pacific Ocean.
(Courtesy NASA)

isolated from the open Pacific Ocean by the Channel Islands, off Santa Barbara to the north (Figure 8.2). Bays, harbors, and fjords have restricted communication with the sea. Therefore, it is not surprising to find that near coasts, seawater temperatures and salinities often change significantly within a few tens or hundreds of kilometers. Near coasts, tidal ranges are also larger and tidal currents stronger than in open-ocean waters.

8.2 COASTAL WATERS

Water characteristics commonly vary substantially in coastal waters. Rivers (Figure 8.1) discharge large amounts of fresh water into coastal waters. Consequently, salinities are generally lowest near river mouths and highest in the centers of open-ocean gyres far from land. Low-surface-salinity waters generally parallel coastlines.

Temperatures in coastal waters are controlled primarily by the incoming energy received from solar radiation. Thus, water temperatures are highest in the mid-latitudes, where insolation is largest, and are lowest in polar regions, which receive little insolation and where sea ice buffers surface waters from temperature changes. (This was discussed in Chapter 3.) Temperatures of coastal waters are also affected by oceanic currents transporting either warm or cold waters and by wind-induced upwelling of cold, subsurface waters. Cold winds off the land chill coastal waters during winter, and sea ice forms in higher-latitude waters and in isolated shallow bays and estuaries.

8.3 COASTAL CURRENTS

Winds and river discharges control coastal currents. Wind-driven coastal currents are highly variable because winds are so changeable. Surface currents respond to changes in winds within a few hours. Storms lasting several days can set up strong currents. (This is one reason why slow-moving, long-lasting winter storms, called northeasters or extratropical cyclones, cause so much damage to beaches and coastal installations.)

Density variations in coastal waters are dominated by salinity changes. Salinity variations in coastal waters persist longer than wind effects. Indeed, winds essentially control density distributions in coastal oceans, which in turn control geostrophic currents.

Remember that the sea surface above low-density waters stands higher than sea surface above high-density waters (Figure 8.3). Thus, low-salinity waters along a coast cause sea surfaces to slope (ignoring temperature effects on density) higher near the coast and lower near the open sea. Water responds to such a sloping surface as it does in the open ocean. As water runs downhill it is deflected by the Coriolis effect and moves along the sloping surface. This causes

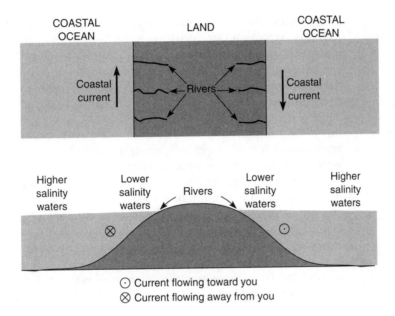

FIGURE 8.3
Sloping sea surfaces and coastal currents resulting from river discharges into coastal waters of the Northern Hemisphere.

coastal currents to flow northward (the seasonally present Davidson Current, for example) along the Oregon-Washington coast and southward along Canada's Atlantic coast (such as the Labrador Current).

Satellite images show how complex these coastal currents are and how they respond to irregular shorelines and rugged continental shelves. It will soon be possible to model and thus predict these currents using knowledge of the winds and river discharges that cause them.

8.4   ESTUARIES AND ESTUARINE CIRCULATION

Estuaries and estuarine circulation patterns are typical of coastal-ocean processes. Estuaries are partially isolated basins (often drowned river valleys) where river waters mix with sea waters. Estuaries also function in other ways. For example, they trap riverborne sediment particles and also support abundant growth of microscopic plants. Estuaries and nearby wetlands serve as nursery grounds for the early stages of many important fishes and shellfishes.

Estuarine circulation (average or nontidal flow) is a mixing system involving two-way water flows; surface waters (less saline, lower density) move seaward whereas landward-moving subsurface seawaters (higher density) flow into estuaries along the bottom (Figure 8.4). Because most coastal waters receive river discharges and are thus lower in salinity than the open ocean, they normally exhibit the estuarine circulation patterns which we now discuss.

Seaward-flowing surface waters flowing over landward-moving seawaters draw them up from below and mix with them, which causes the surface waters to become more saline as they move seaward. The resulting brackish waters, being less dense than open-ocean seawater, remain near the surface. Such mixing is essentially a one-way flow for bottom waters—upward into the surface layers—which then requires landward flows along the bottom to replenish seawaters that have flowed upward (Figure 8.4).

To see how this mixing works, let's compare salinities of surface and bottom waters and also the amount of seawater flowing in along the bottom to the amount of river water discharged. Mixing equal volumes of river water (S = 0) and seawater (S = 35) gives us a mixture of intermediate salinity, 17.5. Two volumes of seawater and one volume of fresh water form brackish waters, with a salinity of 23.3.

By the time surface waters leave estuaries they usually have salinities of around 30. This means that the amount of water moving in along the bottom is 10 to 20 times greater than the river's freshwater discharges.

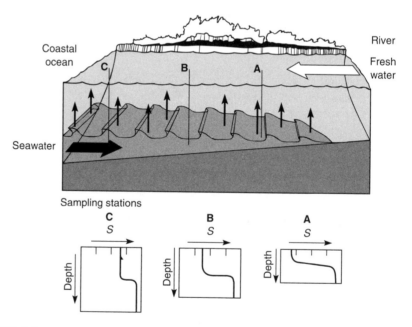

FIGURE 8.4
Vertical water circulation in a simple, salt-wedge estuary. Variations in salinity with depth at three stations (A, B, and C) are shown in the lower part of the figure. Note the pronounced two-layer structure.
[After D. W. Pritchard, "Estuarine Circulation Patterns," *American Society of Civil Engineers, Proceedings* 81 (1955), Separate 717]

Where river flows are large and tidal mixing is small, the two water layers remain quite distinct, forming **salt-wedge estuaries.** For example, during peak discharge of the Columbia River (Oregon, Washington), its estuary has a salt-wedge type of circulation (Figure 8.4), with the tip of the salt wedge extending only a few kilometers into the estuary. At such times, much of the mixing of fresh and salt waters occurs just outside the estuary. During low-discharge periods, salt water occurs much farther up into the estuary, and more mixing occurs there.

Where tidal currents are strong and river discharges small, mixing is vigorous, causing **moderately stratified estuaries** (Figure 8.5). In such estuarine systems, boundaries between surface waters and subsurface layers are diffuse. Chesapeake Bay and its tributaries are moderately stratified estuaries.

An extreme case is the Amazon River (Brazil), where freshwater discharges are so large that salt water is undetectable inside the river mouth. There, mixing of oceanic and fresh waters occurs entirely on the continental shelf.

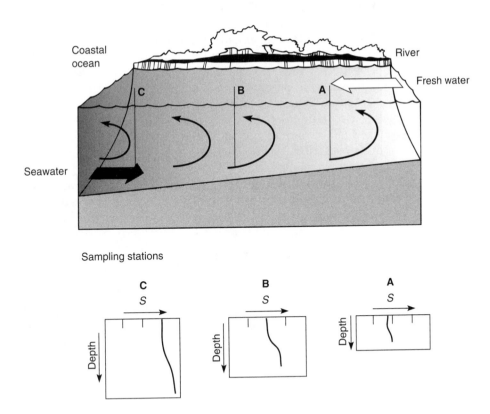

FIGURE 8.5
Variation of salinity with depth at three stations (A, B, and C) in a moderately stratified estuary. Arrows indicate net (nontidal) flows. Note that the two-layered flow here is not as pronounced as in a salt-wedge estuary.
[After D. W. Pritchard, *American Society of Civil Engineers, Proceedings* 81 (1955), Separate 717]

The Coriolis effect affects surface-water movements in large estuaries by deflecting outward-flowing, low-salinity surface waters to the right (in the Northern Hemisphere). These waters flow out along the right side of the estuary (looking seaward), whereas high-salinity waters flow in on the left side (looking seaward). These flow patterns are reversed in the Southern Hemisphere.

Estuarinelike circulation also occurs in coastal- and open-ocean areas that receive more fresh water than they lose through evaporation. For instance, large parts of the northern Indian, North Pacific, and Arctic oceans have estuarinelike circulation patterns.

Off desert coasts, which receive little fresh water and lose much by evaporation, circulation patterns are opposite to those in estuaries. Landward-flowing surface waters replace waters lost through evaporation; seaward-flowing bottom waters carry high-salinity waters resulting from intensive evaporation.

8.5   PARTIALLY ISOLATED BASINS

Land restricts water movements in estuaries, inland seas, and large lakes. Water bodies may also be isolated from the adjacent ocean by sills, which are elevated parts of the sea floor that partially separate basins. These sills restrict bottom-water movements, partially isolating the deeper parts of the basin.

In basins where surface waters are less saline than deeper waters, estuarinelike circulations result. Such circulations are restricted almost entirely to waters shallower than the sill depth (Figure 8.6). Water circulation in silled basins is controlled by the amounts of fresh water flowing in, by winds, and by tidal currents.

Circulation of waters below sill depth is sluggish and predictable. Because of the halocline, bottom waters may be effectively isolated and little affected by surface currents or winds. Water densities outside the basin entrance near sill

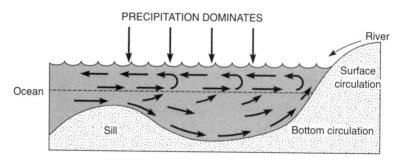

FIGURE 8.6
Generalized water circulation in a silled basin where runoff and precipitation exceed evaporation. Note that the current patterns here (an estuarine circulation) are opposite those in Figure 8.7.

depth affect deep circulation inside the basin. If waters at sill depth outside the basin are appreciably denser than those in the basin, these denser waters flow into the basin, replacing waters that previously occupied deeper parts of the basin. If waters at sill depth outside the basin are less dense than those inside, the bottom waters are not replaced. In many silled and partially isolated basins, such as the Baltic Sea between Sweden and northern Europe, bottom waters are replaced infrequently.

Frequent replacement of bottom waters is necessary to maintain normal marine organisms in deep basins. If bottom waters are not renewed, their dissolved oxygen is consumed by respiration of animals and bacterial decay of organic matter from the surface layers. When the dissolved oxygen is gone, only those organisms that do not need dissolved oxygen for their metabolic processes can survive (these are called **anaerobes**).

Anaerobic organisms obtain needed oxygen by breaking down sulfate ions (SO_4)in seawater and releasing hydrogen sulfide (H_2S), which is toxic to most organisms. Because of these conditions, the bottoms of many stagnant basins are nearly devoid of life, except for bacteria and organisms that tolerate hydrogen sulfide.

The Black Sea is an example of an almost totally isolated basin with an estuarine circulation. Because of its restricted communication with the eastern Mediterranean Sea (through the narrow and shallow Bosporus in Turkey), Black Sea bottom waters are almost completely isolated. Flows through the Bosporus would require 2500 years to replace all waters below a depth of 30 m (100 ft) depth, and hydrogen sulfide occurs below 200 m (700 ft). Thus, most organisms requiring oxygen (called **aerobes**) in the Black Sea are restricted to oxygenated surface waters.

Open-ocean waters outside silled basins usually contain dissolved oxygen. If the basin's deeper waters are replaced frequently, the dissolved oxygen will not be completely used up. The critical factors are the rate at which the bottom waters are replaced, their initial dissolved oxygen contents, and the rate of oxygen utilization in waters below sill depth.

Some **fjords** (flooded deep valleys cut by glaciers) have stagnant or nearly stagnant bottom waters containing hydrogen sulfide. In other silled marine basins, bottom waters are replaced frequently and no stagnation occurs. For example, waters in Puget Sound in Washington (a silled, fjordlike system) are well oxygenated at all depths, because strong tidal flows over sills thoroughly mix the waters. Only in the most isolated parts are dissolved-oxygen concentrations in bottom waters significantly depleted.

Not all basins have an excess of fresh water. In the Mediterranean and Red seas and in the Arabian Gulf, more water is evaporated from the sea surface than is added through river discharge or precipitation. Consequently, surface waters become more saline (Figure 8.7). Surface-water salinities at the northern ends of the Red Sea and Arabian Gulf exceed 40; salinities of surface waters at the eastern end of the Mediterranean Sea exceed 39.

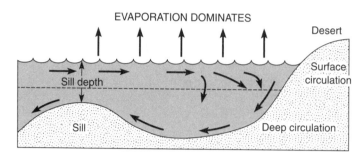

FIGURE 8.7
Generalized water circulation in a silled basin where evaporation exceeds precipitation and runoff.

Just as we saw in desert shelf regions, basins where evaporation exceeds precipitation and river runoff have circulation patterns opposite to estuarine circulation. In evaporating basins, surface waters become denser than those below, and they sink. At the basin entrance, surface waters flow into the basin to replace waters lost by evaporation. Subsurface outflows of warm, salty waters remove the salts left behind by the evaporated surface waters.

In both the Mediterranean and Red seas, the warm, saline waters are cooled in winter, becoming denser. Then they sink below the surface and flow out through narrow straits and into the nearby open ocean. These subsurface water masses can be identified by their distinctive temperatures and salinities as they move along surfaces of constant density across many thousands of kilometers. Again, we are still learning about the processes by which such water masses form and maintain their identities.

8.6 LARGE LAKES

Although they occupy basins that often are thousands of kilometers from the ocean, large lakes share many processes with coastal-ocean areas. And they face many common problems, because most large lakes have densely populated shores.

North America's Great Lakes (Figure 8.8) form a midcontinent maritime environment. These lakes contain about one-fifth of Earth's fresh water, and the lands draining into them are home to 37 million people in the United States and Canada. Like coastal oceans, they are highly vulnerable to abuse and overexploitation of living resources.

The five Great Lakes, lying along the U.S.–Canadian border, are about 3800 km (2300 mi) from the Atlantic Ocean. Lake waters take up to two centuries to reach the ocean, eventually flowing down the St. Lawrence River and through its estuary into the North Atlantic. During the last ice age, glaciers blocked various lake outflows at different times. Then their waters reached the ocean by flowing through several different river systems, including the Hudson and Mississippi rivers. Indeed, these changes in the large fresh water discharges into the North Atlantic may have caused some of the abrupt climatic changes observed in the Northern Hemisphere as the continental-sized ice sheets melted and retreated.

The modern Great Lakes and their connecting rivers have been important transportation routes, first used by native peoples and later by European colonists seeking minerals, timber, and furs. The lakes continue to serve as major transportation routes. The lakes also receive large amounts of wastes discharged from the many large cities and industrial facilities around their shores. Cutting of forests, extensive agriculture, and wetland destruction have altered the original lake environment. Overfishing and introduction of new species have greatly changed the character of aquatic life in the lakes.

Effors to clean up the many waste discharges have begun to alleviate problems in individual lakes. Elimination of phosphate from detergents has helped the

FIGURE 8.8
The North American Great Lakes and their drainage basin.

lakes avoid the excessive production of phytoplankton they previously underwent, which had caused anoxic conditions in bottom waters in Lake Erie, the
shallowest and most vulnerable of the lakes. Moreover, bans on the use of certain
insecticides, such as DDT, have helped fish-eating bird populations recover from
the effects of the pesticides that caused thinning of bird-egg shells.

8.7   COASTS

Coasts, where land meets ocean, are home to about half of the world's population. Thus, coasts are both familiar and important to us. We divide coastlines
into two types. One type is formed by marine processes, such as wave erosion,
sediment deposition, and the effects of marine plants and animals (Figure 8.1).
The other type is shaped primarily by processes acting on land (Figure 8.2), such
as volcanic coastlines formed by volcanic eruptions, or uplifting or sinking of
land caused by mountain building (Figure 8.2).

First we consider coastlines shaped by marine processes. Coastlines shaped
by sediment deposition are common (Figure 8.1), occurring along the southern
Atlantic and Gulf coasts of North America. The sediments come from river discharges or from erosion of cliffs and are deposited by currents moving along
shores, often forming **barrier islands** (low sand islands separating shallow
lagoons from the coastal ocean). Marine processes smooth shorelines. In the
process, they isolate indentations and form **lagoons** (elongate, shallow water
bodies parallel to the shoreline). They also partially isolate estuaries.

The rocky shorelines of New England and Atlantic Canada (Figure 8.9)
were formed by continental glaciers that scoured the land during the last ice age,
removing the soils. After the glaciers retreated, the sea rose to its present level
only a few thousand years ago, which is too short a time for waves to have
eroded and smoothed these rocky shores or to have formed barrier islands. Thus,
recently glaciated coastlines have an abundance of small harbors and inlets,
which are favored by sailors and recreational boaters (Figure 8.9).

The rocky shorelines of North America's Pacific coast are caused by continued mountain building raising the land. In most areas of active mountain building, ocean processes cannot keep up with the land's vertical movements. In such
areas, beaches are rare, occurring only near mouths of rivers, which bring sand
to the coast.

Another example of shorelines formed by land processes is **deltas** (low-
lying marshes, wetlands, and swamps formed by sediment deposition, usually at
river mouths). Deltas are normally covered by water-loving, salt-tolerant plants,
which trap sediment, accelerating the building process; the Mississippi River
delta was formed by this process. Since riverborne sediment supplies have been
reduced or cut off (by dams built on the rivers and by soil-conservation activities), wave action and rising sea level have been eroding the low-lying delta
lands. Alteration of Mississippi delta wetlands during oil and gas exploration and
regulation of river flow to prevent flooding have further reduced supplies of sed-

FIGURE 8.9
The rocky coastline of Atlantic Canada has not been exposed to marine processes long
enough to erase the effects of the glaciers that covered the area until about 18,000 years ago.
(Courtesy Newfoundland and Labrador Department of Development and Tourism)

iments to build up the delta surface, which continually sinks. Where sediment
supplies are adequate, wetlands can build upward fast enough to keep up with
rising sea levels or even extend the delta seaward.

8.8 SEA LEVEL

Sea level is rising about 1.5 mm (0.06 in.) per year, much slower than it did
soon after the last continental ice sheets began retreating (Figure 8.10). Most
(about two-thirds) of this rise is caused by the warming and resulting thermal
expansion of ocean surface layers. The remaining one-third is caused by human
withdrawals of waters from rocks on land and from landlocked lakes and basins,
which flow into the ocean. Russia's Aral Sea is an example of water withdrawal.
There, fresh water discharges to the sea were virtually cut off to irrigate cotton
fields. Consequently, the Aral Sea is now about half its original size, and continues to shrink.

FIGURE 8.10
Sea-level changes during the past 18,000 years, initially caused by melting of the continental ice sheets. Recently, the continued sea-level increase has been ascribed to warming of the ocean's surface layer (accounting for about 1 mm, or 0.04 in., per year) and to diversions of fresh waters from inland seas (about 0.5 mm, or 0.02 in., per year)

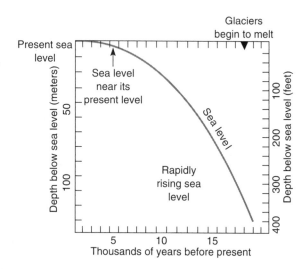

FIGURE 8.11
Rising sea level causes beaches to move landward.

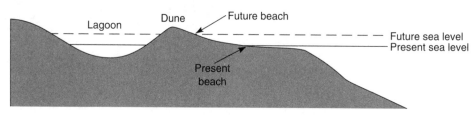

As sea level rises (Figure 8.10), beaches move landward (Figure 8.11), usually slowly but sometimes quite rapidly during slow-moving storms, when beaches may be severely eroded. Such landward-moving shorelines cause erosion of beach fronts and sediment deposition in lagoons behind the beaches. As beaches gradually move landward, previously buried lagoonal deposits are often exposed and eroded.

Since the last glacial retreat began (about 18,000 years ago), beaches have moved many kilometers across continental shelves, reworking sediment deposits as they go. On many continental shelves, old beach features are still recognizable on the submerged bottom. Such features are especially prominent where sea level has remained relatively constant for long periods of time.

If Earth's climate does warm appreciably, sea level will rise faster. The exact amount of potential sea-level rise is unknown, just as the amount of possible global warming is unknown. Estimates of possible sea-level rises with anticipated warming of global climates over the next century (through the year 2100) are:

Present rate: +15 cm (6 in.)
Global warming: low estimate, +50 cm (20 in.)
high estimate, +200 cm (80 in.)

Because most coastal plains are low-lying, each centimeter rise in sea level corresponds to shoreline retreats (Figure 8.11) of meters or even kilometers in some places. A 2-m (6.5-ft) rise in sea level would cause substantial flooding in Florida as well as in many other coastal states.

Many beach homes and other coastal installations are destroyed each year during storms. To protect property against rising sea level and storms, seawalls and other expensive structures are built. If sea-level rises accelerate, expenditures for such protective measures will greatly increase.

8.9   RIP CURRENTS

Near beaches, incoming waves from the open ocean interact with shallow bottoms, causing wave-induced currents nearshore, known as **rip currents** (Figure 8.12). We previously discussed how such interactions cause waves to break in the surf zone. Waves also cause waters to move toward beaches. This, in turn, results in return flows away from the beaches. These return flows form narrow (a few tens of meters across), jetlike currents. These rip currents flow through

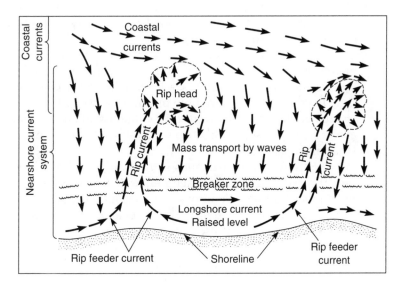

FIGURE 8.12
Waves breaking on beaches induce both longshore currents and rip currents.
[After F. P. Shepard, *The Earth Beneath the Sea*, rev. ed. (Baltimore: Johns Hopkins Press, 1967), p. 18]

the surf zone and up to 1 km (0.6 mi) or more from the beach before the currents are no longer recognizable, blending into the coastal currents.

8.10   WETLANDS

Wetlands (Figure 8.13), also called **salt marshes,** are low-lying, protected coastal areas where fine-grained sediments accumulate. (We will discuss this further in the next chapter.) The upper surfaces of wetlands are generally flat (they are called tidal flats), except where they are cut by meandering tidal creeks through which seawaters flow in to cover the marshes and then drain off during low tides. (There are also wetlands covered by fresh waters, which we will not discuss here.)

Marine wetlands are normally covered by water-loving, salt-tolerant plants. At low tide along the U.S. Atlantic and Gulf coasts, they usually look like grassy meadows. In tropical areas, such as South Florida and Puerto Rico, marine wetlands support extensive growths of salt-tolerant plants called **mangroves** (Figure 8.13). These treelike plants have extensive root systems, which form dense thickets that shelter both marine and land animals. Indeed, some organisms are especially adapted for survival in the mangrove environment. Mangrove oysters attach themselves to roots and branches that are exposed at low tide.

Regardless of the type of plant cover, wetlands are extremely productive. Their production of organic matter equals that of highly productive corn fields. Furthermore, wetlands provide food-rich nursery grounds for many coastal and estuarine organisms.

Roots of wetland plants, both grasses and trees, trap sediment and organic matter. If sea level is rising and the sediment supply is adequate, wetlands build up fast enough to remain at sea level. If sea-level rise is too rapid, marshes are usually eroded as they are now on the Mississippi River delta. If sea level is not changing, wetland plants can trap enough sediment to build the marsh surface above sea level. When this happens, the wetland is replaced by low-lying grasslands or tropical forests.

For many years, wetlands were thought to have little value and were frequently "reclaimed" and turned into agricultural lands or developed for housing or commercial uses. We now know that such marshes are necessary for support of coastal marine ecosystems. Thus, wetlands are now frequently the subjects of legal battles between those wishing to preserve them and those wishing to use them for other purposes. Since 1776, more than half of the United States' original acreage of wetlands has been filled and put to other uses. Now, various organizations are seeking to restore damaged wetlands and even to build new ones. Although wetland systems appear to be simple, it is difficult to build new systems that function as well as undisturbed ones.

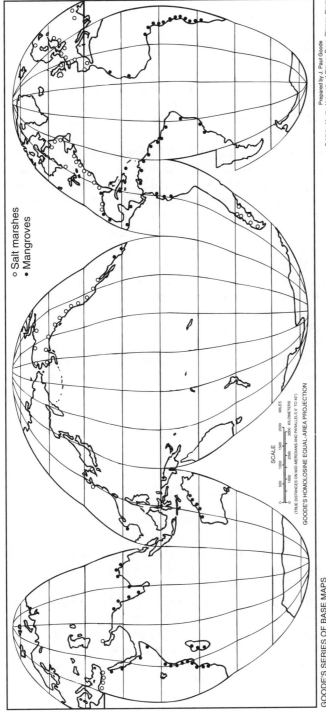

○ Salt marshes
● Mangroves

SCALE

MILES
0   500   1000   1500   2000
0 1000 2000 3000 KILOMETERS

(TRUE DISTANCES ON MID-MERIDIANS AND PARALLELS 0° TO 40°)

GOODE'S HOMOLOSINE EQUAL-AREA PROJECTION

Prepared by J. Paul Goode
Published by the University of Chicago Press, Chicago, Illinois
Copyright 1917 by the University of Chicago

GOODE'S SERIES OF BASE MAPS
HENRY M. LEPPARD, EDITOR

FIGURE 8.13

Wetlands are common features along the world's coastlines. Salt marshes occur on temperate coasts, and mangroves occur on tropical ones.

8.11   GIANT KELP

Giant kelp is a brown alga and the largest marine plant, growing up to 30 m (100 ft) in a single year and reaching 60 m (200 feet) over its life span of seven years or more. It has been called the "sequoia of the sea" because of its large size and relatively long life.

Like all plants, kelp depends on sunlight and thus grows only in relatively shallow waters 6 to 25 m (20 to 80 ft) deep. It grows in the cool offshore waters of western North and South America, Australia, and New Zealand, and around Antarctic islands. These plants need hard bottoms to which to attach themselves, moderate waves, and plenty of nutrients. Under these conditions, they form dense growths, much like forests on land. Buoyant floats on the broad stipes (flattened bladelike structures) keep them near the surface, where they receive the sunlight they need. Holdfasts anchor kelp plants to the bottom, and the stipes grow upward, giving the plants a conical shape. Kelp plants reproduce by releasing spores, which are carried by currents until they reach suitable locations where they germinate, forming new plants.

Dense kelp forests shelter many small organisms, and kelp fronds are eaten by snails, sea urchins, and others. Sea otters living among the kelp eat sea urchins, thus keeping their populations in check. In the nineteenth century, sea otters were heavily hunted (for their fur), almost to extinction. Consequently, sea urchins flourished, and their grazing nearly wiped out kelp forests. Kelp is also exploited by humans. Plant tops are harvested for use in foods and for industrial purposes.

8.12   CORAL REEFS

Coral reefs (Figure 8.14) are large, wave-resistant structures built by carbonate-secreting organisms, primarily colonial corals and calcareous algae. Corals are actually small, colonial animals (Figure 8.15) that form larger structures, each characteristic of a particular type of coral, which are conspicuous on reefs. Calcareous algae are one-celled plants that coat and bind together coral skeletons, forming large, sturdy, cavernous reef structures.

Although corals are animals and feed by capturing floating plants and animals, they also obtain energy from dinoflagellates, called **zooxanthellae,** which live in their tissues. These zooxanthellae give corals extra food, permitting them to grow more rapidly than they could otherwise; without them, corals could not flourish. The zooxanthellae give corals their characteristic bright colors. Thus, corals, although they are actually animals, require essentially the same growing conditions as plants, and require sunlight because of their zooxanthellae.

If sea level rises, coral reefs must grow upward to maintain themselves near the surface. But if sea level rises too rapidly, corals may not be able to grow fast enough to keep up; they die when too deeply submerged. Hundreds of sub-

FIGURE 8.14
Corals are conspicuous components of reef surfaces. Encrusting algae cover the spaces
between coral heads, binding them together. Note the many fishes living around the reef.
(Copyright Great Barrier Reef Marine Park Authority. Used by permission)

merged Pacific seamounts are capped by drowned reefs. Submergence of their
platforms also requires corals to grow upward to compensate for the subsidence.

Coral reefs are found primarily in tropical Indian and Pacific ocean waters;
a few occur in the Caribbean (Figure 8.16). Most reef-building corals live in
warm waters with average annual temperatures between 23 and 25°C (73 and
77°F). Most corals cannot tolerate prolonged exposures to low temperatures or
to large temperature changes.

Reefs also provide nearly closed systems where nutrients are produced,
conserved, and recycled locally, while surrounded by nutrient-poor open-ocean
waters. These nutrients support growth of phytoplankton, which feed benthic
organisms living on the reef as well as fish and plankton living there. Consequently, coral reefs are oases of marine life surrounded by areas of low-produc-
tivity open-ocean waters.

FIGURE 8.15
Drawing of a simple coral
polyp.

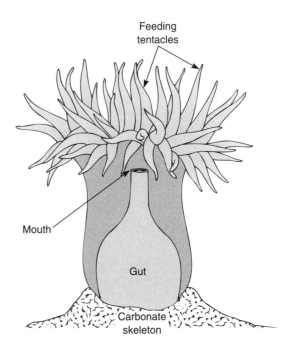

Feeding
tentacles

Mouth

Gut

Carbonate
skeleton

In recent years, coral reefs have been increasingly threatened with destruction caused by ships running aground on them and by dredging for navigational channels. Oil spills have also taken their toll. But the most serious threats have come from population growth along nearby shores. Construction of houses releases sediment, which can smother corals and prevent them from getting enough sunlight. Later, discharges of nutrients in sewage can also destroy the reefs, by supporting algae that grow on the corals. In the late 1980s and early 1990s, large coral reef areas underwent "bleaching episodes" when corals on many reefs lost their zooxanthellae, apparently as a result of stress. Some reefs have recovered, but not all. In short, large areas of coral reefs are now seriously threatened, and these threatened areas may be spreading. Efforts are now underway to protect reefs against further degradation.

8.13   OFFSHORE FACILITIES

Increasing coastal populations need space to build factories, transportation facilities (ports and airports), and recreational facilities. This is especially true in densely populated coastal regions and small island nations. Both Japan and the Netherlands have long experienced such pressures. In both countries, new facilities have been built on nearby shallow sea floors.

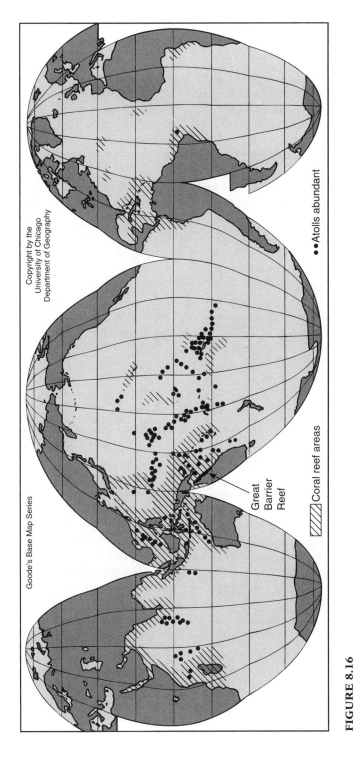

Goode's Base Map Series

Copyright by the
University of Chicago
Department of Geography

Great
Barrier
Reef

Coral reef areas

•• Atolls abundant

FIGURE 8.16

Distribution of coral reefs and atolls. Note that they occur only in tropical and subtropical waters. They are also missing near the mouths of major rivers and where coastal upwelling makes the waters too cool for corals to survive.

These pressures to expand seaward are not new. Between 1200 and 1950, the Dutch "reclaimed" about 6300 km² (2400 mi²) for agricultural purposes. Large, shallow, continental-shelf areas were enclosed by dikes (low earthen dams) to form agricultural fields. Some projects were built on monumental scales. In the 1930s, the Zuider Zee, a large shallow embayment, was cut off by a dike, changing a brackish estuary into a freshwater lake. Then agricultural lands were reclaimed from former bay bottoms.

More recently, the Dutch have begun removing dikes to return these lands to their original state. One purpose is to provide more protection against flooding caused by storms. Another purpose is to rebuild populations of marine organisms, which had been devastated by loss of living and breeding areas.

Offshore facilities have been built for storage of wastes, especially polluted materials dredged from industrialized harbors. Other offshore facilities include power-generating facilities, port facilities for deep-draft tankers and other vessels, and even offshore islands for industral plants and airports. Such developments are most common in Japan. Offshore platforms for producing and storing oil are common in Gulf Coast waters and in the North Sea. Such facilities make it possible to produce offshore oil and gas fields.

QUESTIONS

1. Describe coastal oceans.
2. Why are coastal-ocean waters so variable in their water temperatures and salinities?
3. Draw a diagram of a Southern Hemisphere geostrophic coastal current caused by river discharge.
4. List some processes that mix coastal-ocean waters. In what parts of coastal oceans is each process most important?
5. Describe the circulation in silled basins where evaporation exceeds precipitation. Give examples.
6. Define estuary. What is the simplest estuarine flow? Describe two different types of estuarine circulation patterns. What causes them to be different?
7. Describe the circulation patterns that result when evaporation exceeds precipitation and river discharge into surface layers.
8. Explain what causes storm surges. Where do they pose the greatest threats to humans?
9. Explain why "reclaimed" lands in the Netherlands are being returned to their original state (shallow sea bottom).
10. Why is the Mississippi River delta now eroding rather than building seaward as it did during the first half of this century?
11. Explain why coral reefs are considered to be under stress and even endangered.
12. Why are wetlands important to marine organisms?
13. Why must corals maintain themselves near the ocean surface?
14. Why do kelp grow only in shallow waters?
15. How did recoveries of sea otter stocks affect kelp forests?

SUPPLEMENTARY READINGS

Books

Bird, E. C. F. *Coasts.* Cambridge: The M.I.T. Press, 1969. Elementary.

Davis, R. A., Jr. *The Evolving Coast.* New York: Scientific American Press, 1994. A modern view of coasts.

Hay, J. *The Great Beach.* New York: Ballantine Books, 1963. Well-written history of Cape Cod beaches.

Articles

Bascom, W. "Beaches." *Scientific American* 203(2):80–97.

Carr, A. P. "The Ever-Changing Sea Level." *Sea Frontiers* 20(2):77–83.

Emiliani, C. "The Great Flood." *Sea Frontiers* 22(5):256–270.

KEY TERMS AND CONCEPTS

Coastal oceans
Scales of coastal-ocean processes
Salinity and temperature variations
Sea ice
Coastal currents
Estuaries
Estuarine circulation
Salt-wedge estuaries
Moderately stratified estuaries
Partially isolated basins
Silled basins
Anaerobes
Fjords
Effects of excessive evaporation
Water masses formed in evaporating basins
Large lakes
Population pressure

Introduced organisms
Coasts
Marine-formed shorelines
Barrier islands
Lagoons
Terrestrially formed shorelines
Deltas
Sea-level rise
Rip currents
Wetlands
Salt marshes
Mangroves
Kelp
Coral reefs
Zooxanthellae
Offshore facilities
Waste disposal

9
Sediments

Most of the ocean bottom is covered by **sediment deposits,** consisting of mineral grains eroded from soils and rock fragments from volcanic eruptions, mixed with shells and bones of marine organisms. In general, sediment deposits are thin or absent on newly formed crust at mid-ocean ridges and in mid-ocean areas. Conversely, such deposits are thickest on old crustal segments, in enclosed basins, and near land. Much of the ocean's sediments lie at the bases of continental slopes, forming the continental rise (Figure 9.1). Ancient sediment deposits, in the ocean and on land, supply the **fossil fuels** (primarily oil and gas) necessary for modern society, as well as many other industrial materials. Moreover, fossils in ocean sediments record Earth's history and the events that shaped it over the past 200 million years. Older events must be studied in more ancient deposits now found on land.

We begin by examining the origins of the particles that make up sediment deposits. Then we consider various transport mechanisms. Later we consider various environments where sediments accumulate and then turn to the record of Earth's history contained in these deposits. Finally, we discuss oil and gas, the most valuable commodities produced from the ocean today.

9.1   ORIGINS OF SEDIMENT PARTICLES

By studying sediment particles and their sources, oceanographers learn about processes and events controlling Earth's climate and about ocean history and oceanic life. Particles in sediment deposits come from three primary sources: lithogenous particles come from rocks; biogenous particles come from marine

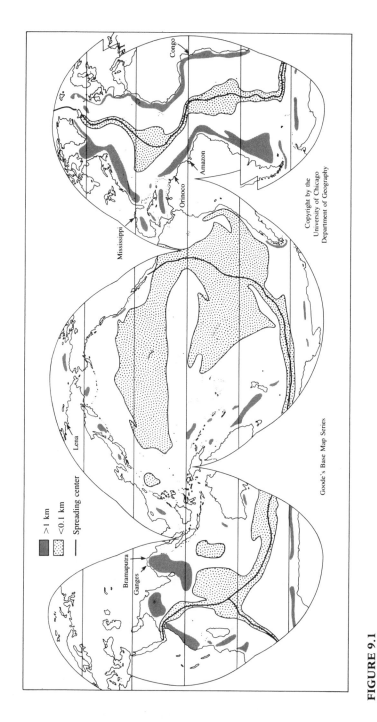

FIGURE 9.1

Thickness of oceanic sediment deposits. Note that deposits are thickest near land, and thinnest on mid-ocean ridges and in the centers of major ocean basins.

[After W. H. Berger, Deep-Sea Sedimentation, in C. A. Burk and C. L. Drake (eds.), *Geology of Continental Margins* (Berlin: Springer-Verlag, 1974]

organisms; and hydrogenous particles are formed by chemical reactions. Let's look at the three types of particles and the processes that form them.

9.2   LITHOGENOUS SEDIMENTS

Lithogenous sediment particles are primarily mineral grains derived from soils formed by physical and chemical breakdown of rocks, called **weathering.** During explosive eruptions, volcanoes also produce lithogenous particles, called **volcanic ash.** Powerful volcanic eruptions that destroyed islands and wiped out civilizations are recorded in ash layers preserved in deep-ocean deposits. Because lithogenous particles come primarily from the land but occur throughout the ocean, we must also examine the processes that transport and deposit these particles—primarily water, winds, and ice.

9.3   SEDIMENT TRANSPORT PROCESSES

Small particles are carried suspended in turbulent water flows. Large grains are dragged along river bottoms by flowing waters and deposited when currents become too weak to carry them farther. Waves and currents on continental shelves sort particles by size, leaving in place the large ones and moving small ones farther from their source.

The largest particles remain nearest their sources—on shores or beaches. Smaller particles are more readily transported seaward and deposited in deeper water on continental shelves and slopes. The smaller particles may be carried by the ocean surface currents or moved by near-bottom currents. The resulting deposits form bands paralleling the coastline.

Most sediment particles generally do not travel far after reaching the ocean. Large particles brought to the ocean by rivers accumulate in estuaries or in deltas near river mouths. Little river-borne sediment escapes deposition on continental shelves or slopes, except near major sediment-transporting rivers such as the Mississippi and the Ganges. There, thick sediment deposits form on the deep-ocean floor, hundreds of kilometers from a river mouth, after turbidity currents carry them many kilometers out onto the deep-ocean floor, which we discuss later.

Virtually every part of the ocean receives **windborne dust** or fine sand particles; indeed, such particles dominate deep-ocean deposits in the centers of major ocean basins, especially around 30°N and 30°S, where deserts are especially common. Mountains barren of vegetation are also major sources of wind-blown dusts, in part because of the ease of eroding their surface deposits and because of their high winds.

The resulting fine-grained deposits accumulate slowly, particle by particle; these are called **pelagic deposits.** Thus, a layer about 1 mm thick forms in 1000 to 10,000 years (a layer 1 in. thick in 25,000 to 250,000 years). The behavior of these particles is like a light snowfall, and the resulting deposits drape over the bottom like a blanket of snow on land, unlike turbidity currents (discussed later), which can completely bury the bottom topography.

Regardless of how they get there, very small particles also sink slowly through the ocean depths and are widely dispersed by currents. Their long transit times also provide ample opportunity for chemical reactions to occur. (We discussed their effect on sea-salt composition in Chapter 3.) For example, iron coatings on particles react with dissolved oxygen in seawater, forming rustlike (iron oxide) coatings. The abundance of such red- or brown-stained grains in **deep-sea muds** accounts for the colors and common names of these deposits—for instance, red clays and brown muds. Colors of deep-sea clays commonly vary from brick reds (derived from Saharan Desert sands) in the Atlantic Ocean to chocolate browns in the Pacific.

Near land, especially near large rivers or in deltas, sediments accumulate rapidly, typically at rates of several meters per thousand years, and are buried too quickly to react fully with dissolved oxygen in the water. Therefore, these grains do not react chemically to form reddish or brownish colors but retain a variety of colors, including green and blue. Large amounts of organic matter in sediment deposits can cause them to be quite dark in color.

Glaciers both supply and transport lithogenous sediments. As glaciers flow, the weight of the ice grinds underlying rock surfaces into fine-grained particles. In addition, glacial ice picks up and carries rocks and boulders of various sizes. When glaciers flow into the ocean and the ice melts, they release unsorted mixtures of mud, sand, and boulders. Such **glacial-marine sediments** cover Antarctic continental shelves (Figure 9.2) and are common in the Arctic Ocean and on nearby deep-ocean bottoms.

As a result of these processes, clays are most common on the deep-ocean floor (Figure 9.2). Sand-sized particles make up less than 10% of the deep-ocean sediments. The coarsest deep-ocean deposits are formed from volcanic ash near volcanoes, where explosive eruptions discharge large volumes of ash.

9.4   TURBIDITY CURRENTS

Near major sediment-transporting rivers, intermittent currents of dense, mud-rich waters, called **turbidity currents,** flow down submarine canyons, carrying materials onto the ocean floor, like a watery, muddy avalanche. Such flows have rarely been observed in the ocean but are well known in lakes and reservoirs. Submarine cable breaks, caused by turbidity currents, are common near mouths of major sediment-transporting rivers, such as the Congo (Zaire) and Ganges (in Bangladesh) rivers.

Turbidity currents transport materials from relatively shallow continental shelves or slopes and deposit them on the deep-ocean bottom. Such deposits are

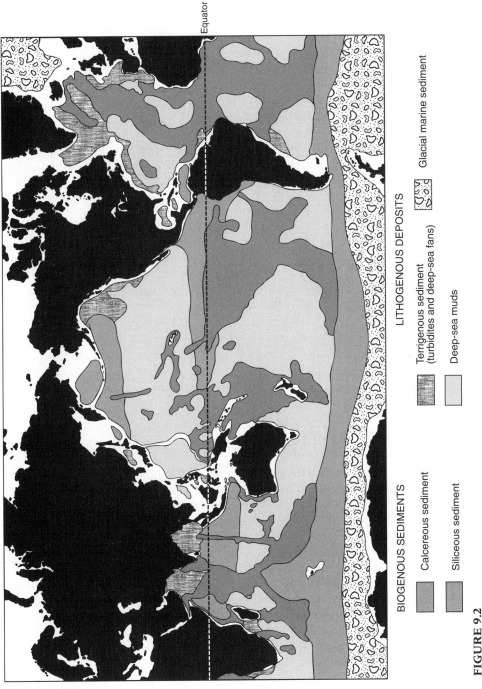

Equator

BIOGENOUS SEDIMENTS

Calcereous sediment

Siliceous sediment

LITHOGENOUS DEPOSITS

Terrigenous sediment
(turbidites and deep-sea fans)

Deep-sea muds

Glacial marine sediment

FIGURE 9.2

Distribution of deep-ocean sediments. Note that siliceous sediments dominate near the
equator and in polar regions. Calcareous sediments dominate mid-ocean ridges.
[After T. A. Davies and D. S. Gorsline, Oceanic Sediments and Sedimentary Processes, in J. P. Riley
and R. Chester (eds.), *Chemical Oceanography*, 2d ed., vol 5. (New York: Academic Press, 1976)]

193

called **turbidites,** or terrigenous sediments (Figure 9.2), meaning that they came from the land.

Although occurring infrequently, turbidity currents transport and deposit sediments over large areas of deep-ocean floor. For instance, a turbidity current triggered by a 1929 earthquake on the Grand Banks south of Newfoundland deposited sediment on the nearby deep-ocean floor over an area 100 km (60 mi) long and 300 km (190 mi) wide. Data on the time elapsed between the time of the earthquake and the time when the flows broke the submarine cables indicate that the currents traveled at speeds of 20 km/hr (12 mi/hr)—about the speed of a slow freight train.

Turbidity currents are much denser than normal seawater, causing them to flow along the ocean bottom, often cutting channels in the ocean bottom, as rivers do on land. Such flows explain the common occurrence of turbidites, deep-water sandy deposits having unusual textures and containing abundant shells and other remains of shallow-water organisms. Submarine canyons near the mouths of major sediment-carrying rivers would soon fill if not emptied by turbidity currents.

Obstructions on the bottom deflect turbidity currents, preventing their flowing onto the deep-ocean floor. This is especially obvious in the Pacific, where submarine ridges formed by island arcs and trenches block flows of near-bottom waters toward the adjacent deep-ocean bottom. In the Atlantic, northern Indian, and Arctic oceans, thick turbidite deposits form the conspicuous continental rises, which we discussed in Chapter 2. Tops of seamounts, submarine ridges, and banks are usually little affected by turbidity currents.

9.5   BIOGENOUS SEDIMENTS

Biogenous sediment particles are produced by organisms, usually by those growing in the photic zone. These particles are primarily from the shells, bones, and teeth of marine organisms. Most organisms are eaten and their remains voided later as relatively large **fecal pellets,** which sink quickly to the bottom. Biogenous sedimentary constituents are further divided into three groups, based on chemical composition: calcareous, siliceous, and phosphatic.

Calcareous constituents (Figure 9.3), primarily calcium carbonate, are most abundant; they consist of carbonate shells (like chalk) of **foraminifera,** coccoliths (platelets secreted by tiny, one-celled algae, Coccolithophoridea), and shells of small, floating snails, the **pteropods.**

Calcareous particles dissolve in the deeper waters of the ocean. Thus, fragile calcareous particles are preserved only at shallow depths under regions where the shell-forming organisms grow in abundance. Some organisms, such as pteropods, occur abundantly in near-surface waters, but their fragile shells are found in sediments only on shallow, seamount tops. At greater depths, the shells dissolve and are not preserved (Figure 9.4).

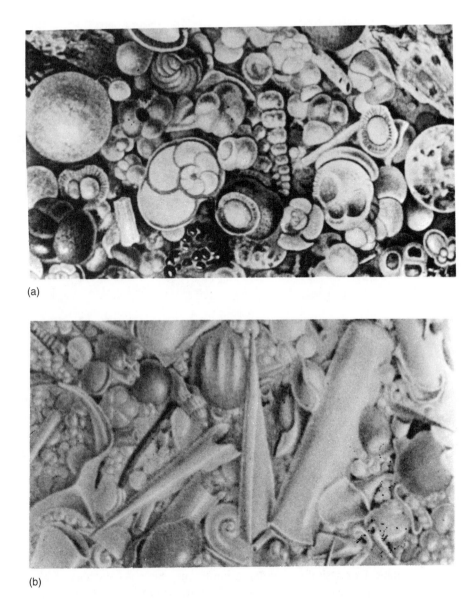

(a)

(b)

FIGURE 9.3
Calcareous deep-sea sediments, containing (a) globigerina (robust-shelled foraminifera) and (b) pteropods (planktonic snails) having delicate, easily dissolved shells.

Robust shells (many foraminifera) survive long enough to reach the bottom. If they are covered by later deposits before being dissolved, they will persist in the sediments. Calcareous muds cover nearly half the deep-ocean bottom and are most abundant on the shallower parts (less than about 4500 m, or

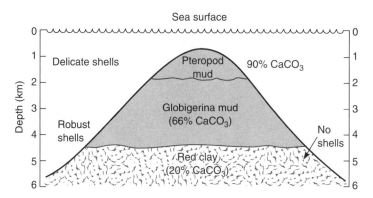

FIGURE 9.4
Depth zonation of calcareous deposits, resulting from dissolution of calcareous con-
stituents in near-bottom waters, leaving red clays, with no calcareous shells.
[After J. Murray and J. Hjort, *Depths of the Ocean* (London: Macmillan, 1912), p. 173]

15,000 ft). They accumulate at rates between 1 and 4 cm (0.4 and 1.6 in.) per
thousand years.

Siliceous constituents (Figure 9.5), second in abundance, consist of opal
(glasslike) shells, secreted by **diatoms** (one-celled algae) and **radiolaria** (one-
celled animals). These deposits form near the equator, where radiolaria grow in
abundance, and in high-latitude bands (Figure 9.2), where diatoms grow in
great abundance (discussed in Chapter 7). Siliceous particles dissolve at all
depths in the ocean, and so siliceous sediments form primarily under areas
highly productive of siliceous organisms, such as radiolaria.

Siliceous muds are most common in the Pacific; diatom-rich muds nearly
surround Antarctica and occur in the North Pacific; radiolarian-rich sediments
dominate the equatorial Pacific. Both ocean areas are highly productive of
marine life (Figure 9.2).

Phosphatic constituents (rich in phosphate) are rare in sediments. They
consist primarily of bones, teeth, and scales of fish, and occur primarily in
deposits on shallow, isolated banks or near coastal areas of high biological productivity. Some nearshore phosphatic deposits may someday be exploited for
their phosphate, which is used in making fertilizers.

Where sediment deposits have not been disturbed by burrowing organisms,
thin annual layers, called **varves,** may be preserved. Such preservation is normally found in isolated basins with anoxic bottom waters (no dissolved oxygen
and thus no sediment-burrowing organisms). Varved deposits are extremely useful for reconstructing past events. For instance, studies of fish scales taken from
individual layers show large decade-scale fluctuations in fish abundances
extending over many centuries. Such studies have been especially informative
along the Pacific coast of the Americas in analyzing fluctuations in abundances
of sardines and anchovies.

(a)

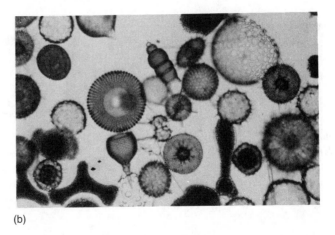

(b)

FIGURE 9.5
Siliceous constituents of deep-sea sediments. Diatoms (a) and radiolaria (b) are about the size of grains of sand.
[Courtesy W. R. Reidel, Scripps Institution of Oceanography]

As biogenous particles sink to the bottom, the most soluble ones dissolve. **Dissolution** occurs as fragments sink and continues while they lie unburied on the bottom. The longer the grains remain in contact with ocean water, the more they dissolve. At first, only the most robust shells remain. Finally, all carbonate shells and fragments are dissolved.

A final factor affecting distributions of biogenous sediment is **dilution**, or mixing with other types of particles. Because biogenous sediment contains (by definition) more than 30% biogenous constituents, it is clear that even large amounts of biogenous particles may be diluted by large amounts of lithogenous sediment and therefore may not form biogenous deposits.

9.6 HYDROGENOUS SEDIMENTS

Hydrogenous sediment constituents are formed by chemical reactions in seawater or within sediment deposits; irregularly shaped **manganese nodules** (Figure 9.6) are the most familiar examples. These sooty black nodules are found in sizes ranging from the size of a small pea to that of a coconut; some even form slabs. Mixtures of iron-manganese minerals precipitated in seawater coat objects on the sea floor, forming roughly concentric bands around rocks or whale's earbones or anything that remains exposed on the ocean bottom for very long times. It seems likely that burrowing organisms turn over nodules so that accumulation can occur on all sides.

Manganese nodules grow extremely slowly, a layer a few hundredths of a millimeter to a millimeter thick forming in a thousand years. Because they accu-

FIGURE 9.6
Potato-sized manganese nodules are abundant on the floor of the South Pacific.
[Courtesy National Science Foundation]

mulate extremely slowly, manganese nodules are abundant on the bottom far from land and where biological productivity is low. Otherwise, lithogenous or biogenous constituents dominate the deposits and bury the nodules.

Manganese nodules occur over much of the deep-ocean floor—especially in the Pacific, where they cover 20 to 50% of the bottom. The central Pacific band of abundant manganese nodules is interrupted by siliceous muds along the equator, which apparently accumulate much more rapidly.

Chemical processes forming manganese nodules are common, but in areas of rapidly accumulating sediments, small (pea-sized or smaller) micronodules form and are buried by rapidly accumulating sediments, instead of forming nodules or slabs at the surface.

Manganese nodules are potential sources of copper, nickel, and cobalt. Metal-rich nodule deposits occur in the Pacific, especially in the North Pacific south of Hawaii. Cobalt-rich manganese crusts occur at intermediate depths (2 to 3 km, or 1 to 2 mi) on rocks around many Pacific islands. But after many years of exploration and development of deep-ocean mining techniques, there is still no commercial nodule production from the ocean floor.

9.7   BEACHES

Beaches are deposits of loose sand and gravel moved by waves and currents along shorelines. They occur near sediment sources, such as eroding cliffs and river mouths. Most beaches have certain features in common. Behind most beaches there are dunes (deposits of wind-blown sand) on low-lying shores or sea cliffs on more rugged ones. Moving toward the water, the berm (the backshore part of the beach) gently rises to a crest, and from there the foreshore slopes seaward. Offshore, there are usually several submerged bars (Figure 9.7).

Waves dominate beach processes. Most sediment transport takes place between the upper limit of wave advance and water depths of 10 to 15 m (30 to 50 ft). Where waves are exceptionally strong, fine particles are removed, leaving coarse sand or gravel on the beach. Fine-grained materials are deposited where there is little wave action (Figure 9.8), such as in protected bays and lagoons behind **barrier islands** (long islands paralleling the shore, built of sand) or **barrier beaches.**

Beaches change seasonally. In winter, most beaches are eroded by short, choppy waves formed in nearby storms. Strong wave-induced currents cut deep channels. Sand removed from beaches is transported to nearby submerged bars. Strong winter winds also blow dry sands from beaches onto the dunes behind them.

In summer, waves are usually low, long-period swells from distant storms, and sand moves from offshore bars to rebuild nearby beaches. Longshore bars fill in and the summer beach profile becomes less steep than during the winter.

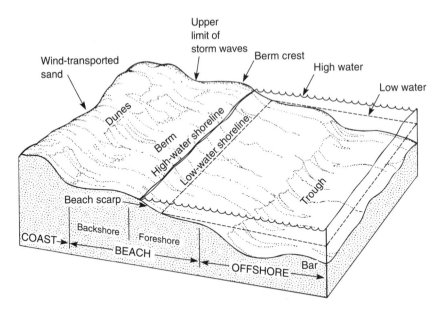

FIGURE 9.7
A beach and adjacent features.

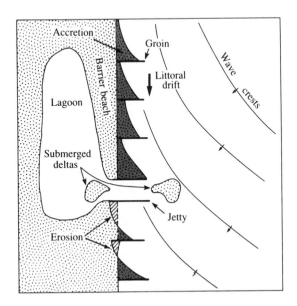

FIGURE 9.8
Waves obliquely approaching shores move sands along beaches; this is called littoral drift. Sand movements are slowed and deflected by groins (low walls) and by jetties (low stone walls), protecting navigation channels. While expensive, neither type of structure is totally effective.

Waves usually approach beaches obliquely, causing **longshore currents** that move sediment along the beach. These sediment movements are called **littoral drift.** Many thousands of tons of sand are moved each year along most beaches. The moving sands are trapped at inlets along barrier beaches and deposited on small submerged deltas or in lagoons behind barrier beaches (Figure 9.8).

9.8   DEEP-OCEAN SEDIMENTS

New instruments and techniques now make it much easier to study processes and discrete events that affect deep-ocean sediments. These studies help scientists to interpret sedimentary features that may have global significance, such as changes in the deep-ocean current patterns.

Western boundary currents, such as the Gulf Stream and the Kuroshio, have considerable strength to depths as great as 2 km (1 mi) below the surface. Where deep currents contact the bottom, they erode sediment deposits, leaving breaks in the sedimentary record. The eroded materials form near-bottom layers of turbid waters, which are moved by deep-ocean currents.

Sediment deposits also record intermittent passages of rings spun off from major boundary currents. These rings, called **abyssal storms,** erode the bottom. They have strong currents lasting for a few days to a few weeks and not only erode the deposits but also form ripple marks like those seen in the bottoms of fast-flowing mountain streams. Areas of such ripple marks occur under the Gulf Stream off Florida and off the Carolinas, where strong currents scour the bottom.

Ripple-marked sediments also occur in areas affected by outflows of large subsurface water masses. Areas off Spain and Portugal are affected by **subsurface outflows** from the Mediterranean, along the bottom of the Strait of Gibraltar. Sediments in the northwestern Atlantic record outflows of dense waters that make up the North Atlantic Deep Water.

By selecting areas that are likely to have been exposed to certain types of intermittent processes, scientists can reconstruct ocean history and conditions as recorded in sedimentary deposits. For instance, studies of pelagic sediments deposited on the tops of oceanic plateaus provide information about events affecting the overlying waters. Sediments at intermediate depths can record **anoxic events** that occurred when ocean circulation was too sluggish to replenish dissolved oxygen in bottom waters. These events are especially important because much of the world's oil and gas deposits apparently formed under anoxic conditions.

9.9   EARTH'S CLIMATE RECORD

In their efforts to determine how the ocean and atmosphere have responded to climatic and other changes in Earth's history, scientists study a variety of particles and features from which indications of Earth's climate can be deduced.

Obviously it would be easiest to use recorded instrument readings, such as readings from barometers and thermometers, but such records go back only about a century. Human history provides written records that go back about 2000 years in Egypt, China, and areas around the Mediterranean. But such records are sparse or nonexistent for most ocean and polar areas.

For most studies of Earth's history, scientists must use less precise records, which must be carefully interpreted. Tree rings are familiar examples of climatic data: thick tree rings indicate wet years, whereas thin rings denote dry years. Comparable records can be obtained from the rings laid down annually by some corals.

Sediment deposits in modern ocean basins are prime sources of data. Sediment deposits on the ocean floor cover the most recent 200 million years. For earlier times, the records are in ancient marine sediments preserved on land, as indicated in Table 9.1.

Sediment deposits in the ocean and those preserved on land contain various features that indicate climatic conditions (these are called **proxy indicators**) up to several billion years ago. Remember that the oldest sediments are 3.8 billion years old and have been altered chemically and physically by various geologic processes. In general, the older the deposit is, the more difficult these proxy indicators are to interpret.

Sediment features are used to estimate sea levels over the past 100 million years, which have been as much as 250 m (820 ft) higher than at present. Also, the detailed chemical compositions (variations in isotopic oxygen content) of carbonate shells of marine organisms are used to determine the seawater temperatures under which the organisms grew. Annual bands laid down by long-lived corals provide detailed records extending back several thousand years.

By understanding how Earth's climate has responded to naturally changing conditions, scientists will be better able to predict the effects of human alteration of the atmosphere. For instance, over much of Earth's history, its climate has oscillated between cold periods (such as the present) and periods when tropical

TABLE 9.1
Sources of Data Indicating Earth's Climatic History

Data Source	Period Covered (yr)
Instrument observations	100
Written history	2,000
Tree rings	14,000
Coral rings	100,000
Glacial ice	300,000
Lake sediments	1 million
Ocean-basin deposits	200 million
Sediment deposits on land	3.8 billion

climates extended well into the polar regions. In short, Earth underwent major climatic changes long before humans began polluting the ocean and the atmosphere. By understanding these naturally occurring changes, we can better evaluate the possible effects of atmospheric and oceanic pollution resulting from human activities.

9.10   GLACIAL OCEAN

During most of its history, Earth's climate has been warm and humid, with little difference in surface temperatures in tropical and polar regions. The past 2 million years, however, have been markedly cooler than much of Earth's history. One way to look at Earth's climatic record is that it has two climatic states: the greenhouse state, when there is little temperature difference between the tropics and the poles; and the "icehouse" state, such as today, when there are large permanent ice sheets near both the North and South poles.

Continental glaciers advanced and retreated at least 30 times over large portions of the Northern Hemisphere continents. Times of glacial advance are **glacial stages;** times of glacial retreat are called **interglacials.** (We are now in an interglacial stage.)

Sea level fell to about 130 m (430 ft) below its present level during glacial periods and at times rose to about 50 m (160 ft) above the present level during interglacial periods. The most recent retreat began about 18,000 years ago. Today's climate (an interglacial period) is warmer than it has been for most of the past 2 million years.

The distributions, abundances, and variations in the chemical compositions of the materials making up the foraminiferal shells in ocean sediments show how glacial advances have affected ocean temperatures (Figure 9.9). From such data, scientists have determined that 18,000 years ago, average sea-surface temperatures were 2 to 3°C (4 to 5°F) lower than they are now; average land temperatures then were about 6.5°C (12°F) lower than today. At the same time, ocean currents were more energetic, and their positions were different. For example, the Gulf Stream went nearly due east from the Carolinas, striking Spain rather than flowing into the Norwegian Sea as it does today. When sea level stood about 85 m (280 ft) lower than it stands today, humans migrated from Siberia to Alaska across the exposed Bering Sea floor to populate the Americas.

Regular changes in various aspects of Earth's orbit around the Sun appear to have been major factors controlling these glacial advances and retreats. This theory, worked out by the Serbian mathematician Milutin Milankovitch (1879–1958), predicts climatic changes recurring every 100,000, 40,000, and 10,000 years. Each area responds differently to the small changes in solar heating caused by the slight changes in Earth's orbit around the Sun. Such changes in Earth's climate have occurred repeatedly over much of Earth's history.

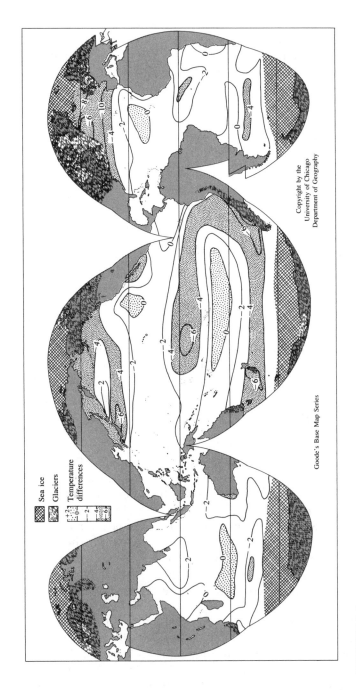

FIGURE 9.9

Changes in sea-surface temperature since the end of the last glacial advance (18,000 years ago) of the ice age. Water temperatures were lower at high latitudes and along the equator, and higher in the centers of the subtropical gyres, than at present.
[After A. McIntyre, CLIMAP, "The Surface of the Ice-Age Ocean," *Science* 191 (1976), p. 1134]

Legend (within figure):
Sea ice
Glaciers
Temperature differences
+2
0
−2
−4
−6

Goode's Base Map Series

Copyright by the
University of Chicago
Department of Geography

9.11   CATASTROPHES AND EXTINCTIONS

Large-scale extinctions of plants and animals, recorded in sediment deposits (Figure 9.10; Appendix 2), indicate that Earth has undergone many catastrophes, some much larger than glacial advances and retreats. Some catastrophes have apparently been caused by asteroids or comets striking Earth, whereas others apparently have resulted from processes originating deep within Earth itself, such as eruptions of huge volumes of volcanic rock in short periods of time, causing large variations in climate.

Meteor impact craters are conspicuous on the surfaces of the Moon, Mars, and Mercury, but not on Earth, which undoubtedly has undergone similar bombardments. On other planets and on the Moon, little happens to impact craters once they are formed—except possibly other meteorite strikes. But on Earth, most meteor impact craters are destroyed by subduction or eroded and later obscured by sediments deposited over them. For example, only four of the predicted 40 to 160 craters larger than 30 km (20 mi) in diameter have been found on Earth. Only one out of 18 craters preserves the deposits formed by the impact. Furthermore, we know little about the timing of events; only eight craters have been dated to within 10 million years.

The continuing search for the cause of the demise of the dinosaurs 65 million years ago began with studies of unusual, thin, sediment layers, which con-

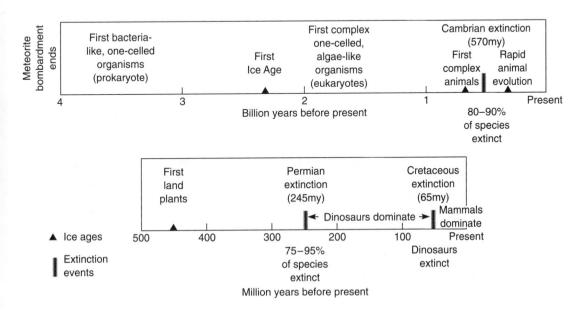

FIGURE 9.10

Some major evolutionary advances in life on Earth and some of the major extinction events recorded in sediment deposits.

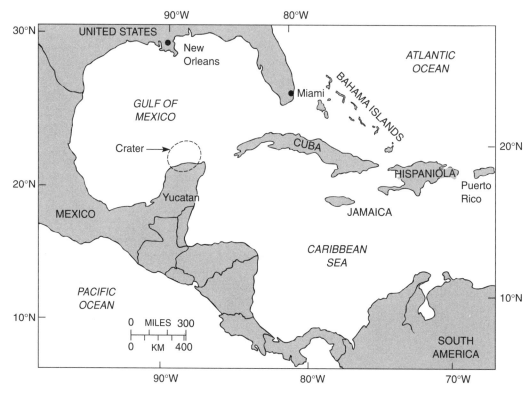

FIGURE 9.11
The large buried structure north of Mexico's Yucatan Peninsula is apparently a crater
formed by a meteorite impact.

tained high concentrations of the element iridium. These iridium-rich layers
were deposited at the same time that dinosaurs and many other plants and animals became extinct.

Iridium, a platinumlike metal, is scarce on Earth's surface but abundant in
asteroids and comets. These iridium-rich sediment layers are found worldwide.
This led to the hypothesis that a mountain-sized extraterrestrial object (about 10
km, or 6 mi, across) struck Earth, causing the extinction of the dinosaurs in a very
short period of time. (The mechanisms causing extinction are still disputed; other
scientists claim that massive volcanic eruptions are a more likely explanation.)

Among the clues thought to indicate possible impact-crater locations are
deposits around the Gulf of Mexico–Caribbean area which apparently have been
disturbed by enormous waves; similar deposits have been found near Haiti and
Cuba. An unusually large structure north of Mexico's Yucatan Peninsula (Figure
9.11) is buried by limestones. It has been partially penetrated by wells drilled for
oil, but further studies will be required to determine if the structure was formed
by a meteor impact or by a gigantic volcanic explosion. Estimates of the struc-

ture's diameter range from 170 to 300 km (100 to 200 mi), making it one of the largest craters in the Solar System.

Impacts from extraterrestrial bodies are only one possible cause of catastrophic extinctions. Extinctions may have also been caused by massive outpourings of lava (much larger than those of single volcanoes). These eruptions apparently were fed by unusually large masses of molten rock, which may have originated at the core/mantle boundary deep in Earth's interior. Such huge masses of molten rock, called **superplumes,** are thought to rise quickly through the mantle to reach Earth's surface. The resulting volcanic eruptions may last only a few million years. Along with the lava, large amounts of carbon dioxide and other gases are released into the atmosphere. These events apparently caused changes in Earth's climate, which could also have led to worldwide extinctions of plants and animals.

Rocks formed by such massive eruptions are well exposed on land. For instance, one volcanic province (formed about 16 to 18 million years ago) is exposed along the Columbia River gorge (Washington and Oregon). Other large volcanic provinces (thought to have formed in much the same way) occur on the ocean bottom and are poorly known; two examples are the Ontong Java Plateau in the Pacific (about 120 million years ago) and the Kerguelen Plateau in the southern Indian Ocean (about 110 million years ago). Formation of these two provinces coincides with the time of a major extinction episode, as shown in the fossil record.

Massive volcanic eruptions may have also contributed to the demise of the dinosaurs. At about the same time as the meteor impact about 65 million years ago, massive volcanic eruptions occurred in India, forming a volcanic province known as the Deccan Traps, now partly submerged on the ocean bottom in the northern Indian Ocean. Climatic changes caused by these eruptions may have reduced the numbers and range of dinosaurs before they became extinct.

Such events have occurred many times in Earth's history. Some of them nearly caused the total extinction of life on Earth. The Permian event, about 245 million years ago, caused the extinction of up to 95% of the species living at that time (Figure 9.10).

Massive eruptions of individual volcanoes (Figure 9.12) also affect Earth's weather worldwide. For instance, the 1991 eruption of Mount Pinatubo in the Philippines was the largest in the last century. Volcanic ash and gases injected into the stratosphere caused unusual weather and cooler climates around the globe for several years. But none of these historically documented events caused major extinctions.

9.12   PETROLEUM

Oil and gas are the most valuable materials now produced from the ocean bottom (Figure 1.12). Furthermore, the most promising areas for future exploration for new oil and gas fields are continental margins, especially continental slopes.

FIGURE 9.12
A volcano off Iceland erupted in 1963, injecting large amounts of volcanic ash and gases
into the atmosphere. Similar volcanic eruptions of Icelandic volcanoes in 1783 caused
widespread cold temperatures and crop losses in Europe. Such bad weather during the
1780s likely contributed to the French Revolution in 1789.
(Courtesy Icelandic Airlines)

Let's see how oil and gas are formed and why the ocean offers the best hunting
ground for new discoveries.

Oil and gas are formed by decomposition of plant materials, usually marine
plants. Most plants are eaten by other organisms, but a small fraction of their
remains reach the ocean bottom and are deposited in sediments. (This organic
matter is what feeds the benthic communities, as we discussed in Chapter 7.)

Where phytoplankton production is exceptionally high in surface waters
and bottom-water circulation is sluggish, dissolved-oxygen concentrations are
depleted when plankton die and decay. Thus fewer benthic organisms can eat
the organic matter once it arrives at the bottom; consequently, more is preserved. Eventually, bacteria partially break down organic matter, forming gas
and oil. Heat and pressure also accelerate these processes, although if temperatures are too high, all the oil is converted into natural gas or into a tarlike substance that is not readily extracted.

Over time, more sediments accumulate on top, compacting those under
them by their weight and expelling water and associated oil in the process.

These expelled fluids move through surrounding sediments until they are trapped in reservoir rocks, typically porous sandstones but sometimes cavernous limestones. Where they have accumulated in sufficient quantities, oil and gas can be commercially extracted.

Most organic matter in sediments is not extracted in this way and remains disseminated in the rocks in which it was formed. Such materials cannot be used as fossil fuels. Also, if the oil and gas fields are too small, it may be commercially unfeasible to explore and drill for them and to build the necessary but expensive production facilities (production platforms and pipelines).

Thus, the factors necessary for the formation of major oil and gas deposits are: (1) thick sediment deposits where organic matter is preserved; (2) porous rocks to hold oil and gas in extractable forms after fluid movements have stopped; (3) an impermeable layer, such as a layer of fine-grained sediment, to keep the oil and gas in what we know as oil or gas fields (otherwise, these fluids would be dispersed throughout the rocks and would not be recoverable); and (4) the passage of many millions of years, to allow these processes to occur naturally. All these conditions occur on continental margins, making them attractive places to hunt for new oil and gas supplies. New production techniques and equipment now make it possible to produce oil and gas from much greater depths than was previously possible.

Large oil and gas fields have been found in the Gulf of Mexico, off southern California, in Alaska, in the Arabian Gulf, and off Nigeria (West Africa). These are extensions of well-known deposits on land. In other areas, such as in the North Sea between England and Norway and in Bass Strait between Australia and Tasmania, all the oil and gas lie under coastal waters. Other promising areas for future oil and gas exploration include the Arctic Ocean continental shelves, Indonesia, and the deeper continental rises.

QUESTIONS

1. What is the most common type of sediment deposit on the deep-ocean floor? Why?
2. Why are the very fine-grained sediments in deep-ocean basins commonly colored red or chocolate brown?
3. In what parts of the ocean are calcareous muds most common? Why?
4. Where are the thickest sediment deposits located in the deep-ocean basins?
5. List some of the resources recovered from the ocean bottom.
6. Describe turbidity currents. List some of the evidence of their existence.
7. List three processes by which lithogenous sediment is transported into the deep ocean. Which is most important? Which is least important? Why?
8. List some of the marine organisms whose shells form biogenous sediments.
9. Why are whales' earbones and sharks' teeth found in manganese nodules?
10. Why are continental margins favorable sites for finding new oil and gas fields?
11. Why are anoxic events important for forming oil and gas deposits?

SUPPLEMENTARY READINGS

Books

Seibold, E., and Berger, W. H. *The Sea Floor: an Introduction to Marine Geology.* Heidelberg, Germany: Springer-Verlag, 1993. A survey of the importance of marine sediment deposits to our understanding of the ocean and its history.

Siever, R. *Sand.* New York: Scientific American Library, 1988. Discussion of techniques and results of studying sands.

Shepard, F. P. *The Earth Beneath the Sea,* rev. ed. Baltimore: The Johns Hopkins Press, 1967. Nontechnical.

Stanley, S. M. *Extinctions.* New York: Scientific American Library, 1987. Popular account of great extinctions in Earth history and their causes.

Articles

Rona, P. A. "Metal Factories of the Deep Sea." *Natural History* 97(1):52–57.

Victory, J. J. "Metals from the Deep Sea." *Sea Frontiers* 19(1):28–33.

KEY TERMS AND CONCEPTS

Sediment deposits
Fossil fuels
Lithogenous sediment particles
Weathering
Volcanic ash
Sediment transport
Wind transport
Pelagic deposits
Deep-sea muds
Glacial-marine sediments
Turbidity currents
Turbidites
Biogenous sediment particles: calcareous, siliceous, and phosphatic
Foraminifera
Pteropods
Diatoms
Radiolaria
Varves
Dissolution
Dilution

Hydrogenous sediments
Manganese nodules
Beach processes
Barrier islands and barrier beaches
Longshore currents
Littoral drift
Abyssal storms
Subsurface outflows
Ocean history
Anoxic events
Climatic changes
Proxy indicators
Sea-level changes
Glacial stages
Interglacials
Extinctions
Meteorite impact craters
Massive volcanic eruptions
Superplumes
Oil and gas formation

Appendix 1
Conversion Factors

LINEAR MEASURE
1 kilometer (km) = 1000 meters (m) = 0.62 mile (mi)
1 meter (m) = 100 centimeters (cm) = 39.4 inches (in.) = 3.28 feet (ft)
1 centimeter (cm) = 10 millimeters (mm) = 0.394 inches (in.)

VOLUME
1 cubic kilometer (km^3) = 0.24 cubic mile (mi^3)

MASS
1 kilogram (kg) = 1000 grams (g) = 2.2 pounds (lb)
1 gram (g) = 0.035 ounce (oz) (28 g = 1 oz)

VELOCITY
1 meter per second (m/s) = 2.2 miles per hour (mi/hr)

TEMPERATURE (DIFFERENTIAL)
1 degree Celsius ($^{\circ}$C) = 1.8 degrees Fahrenheit ($^{\circ}$F)
1 degree Fahrenheit ($^{\circ}$F) = 0.55 degree Celsius ($^{\circ}$C)

TEMPERATURE CONVERSIONS

$^{\circ}$C	$^{\circ}$F
0	32
10	50
20	68
30	86
40	104
100	212

Appendix 2

Important Events in the History of the Solar System and Earth

(Times are approximate; by = billions of years ago; my = millions of years ago)

UNIVERSE FORMS—about 15 by (exact age uncertain)
GALAXIES FORM AND SUPERNOVAS OCCUR—10–14 by
    Supernovas give rise to new stars
    Galaxies evolve by absorbing smaller galaxies
SOLAR SYSTEM FORMS—5 by
    Supernova forms materials incorporated into Sun—4.6 by
    Planets form around Sun—4.45 by
    Moon and Mercury cool, ceasing crustal movements—3 by
    Mars cools, ceasing crustal movements—1 by
LIFE BEGINS ON EARTH—3.8 by
    First prokaryotic cell—3.8 by
    Photosynthesis begins—3(?) by
    First ice ages—2.3 by
EUKARYOTES
    First eukaryotic cell—2.0 by
    Heterotrophy develops—1.0 by
    Sexual reproduction begins—1.0 by
    First multicellular animals—700 my
CAMBRIAN EXTINCTIONS—570 my
    80–90% of species eliminated
PLANTS AND ANIMALS
    Shells developed by trilobites, clams, and snails—540 my
    Vertebrate animals—510 my
ORDOVICIAN CATASTROPHE—440 my
    Jawed fishes appear—425 my
    Finned fishes appear—415 my
    Insects appear—395 my
    Lunged fishes appear—380 my

DEVONIAN CATASTROPHE—370 my
  First trees—370 my
  Amphibians go ashore—370 my
  Winged insects—330 my
  Reptiles and their eggs survive on land—313 my
  Warm-blooded reptiles—256 my
PERMIAN CATASTROPHE—245 my (75–95% of species extinct)
  Dinosaurs and flowers appear—235 my
  Pangea complete—220 my
  Formation of Atlantic Ocean begins—210 my
  Birds appear—150 my
CRETACEOUS EXTINCTIONS—65 my
  Early whales appear (mammals return to sea)—55 my
EOCENE CATASTROPHE—37 my
  Whales become largest marine animals—25 my
  Seals return to ocean—25 my
MIOCENE CATASTROPHE—15 my
  Ice ages begin—3.3 my
  First humans—2.6 my
PLEISTOCENE CATASTROPHE—730,000 years ago
  Last glacial retreat begins—18,000 years ago

Glossary

Abyssal plain   a nearly flat portion of deep-ocean floor, built primarily of sediments from land deposited by turbidity currents

Abyssal storm   a short period of strong near-bottom currents, often accompanying the passage of a ring

Abyssal zone   ocean bottom between 4 and 6 km (2.5 and 3.7 mi) deep

Accuracy   the difference between a measurement and the true value

Acid   a hydrogen-containing compound that is usually corrosive

Acid rain   rain containing acid formed from sulfur and nitrogen compounds released by burning of fossil fuels

Acoustic tomography   a technique of measuring changes in sound velocity between transmitters and receivers to map water-mass distributions and their changes with time in the deep ocean

Adsorption   adhesion of a thin film of a solid or liquid to a solid surface

Advection   transport by the motion of a liquid or gas

Aerobic   occurring where oxygen is present

Aggregate   two or more different kinds of minerals or particles combined into one particle

Air pollution   contamination of the atmosphere by human-made chemicals

Albedo   fraction of sunlight or radiative heat reflected

Algae   one-celled plants, usually containing chlorophyll, which gives them a green color; may have other pigments resulting in other colors

Alkalinity   the extent to which a compound is alkaline (basic) and can neutralize acids by combining with them

Altimeter   radar mounted on an Earth-orbiting satellite to measure distances between the satellite and Earth's surface; maps ocean-surface elevations and depressions, which permits determinations of directions and speeds of surface currents; ocean-surface roughness can also be measured to determine average wave height

Anaerobic   occurring where no oxygen is present

Anoxia   absence of dissolved oxygen in water

Aphotic zone   ocean zone in which no light is present

Aquaculture   commercial cultivation of aquatic organisms, both plants and animals

Arc   a string of islands, usually volcanic, situated behind a trench and above a subducting zone

Arctic Ocean   nearly land-locked ocean basin bordered by North America, Europe, and Asia

Asteroid   a body of rock larger than a meteor but smaller than a planet, orbiting the Sun

Asthenosphere   a plastic layer (possibly containing a small fraction of molten rock) of the upper mantle (from about 100 to 400 km, or 60 to 250 mi, deep) on which the lithosphere floats

Atoll   a ring-shaped coral reef structure rising from deeper water

Autotroph   an organism that can produce its own food

Bacteria   one-celled organisms lacking true nuclei

Baleen   hornlike material growing from upper jaws of certain (baleen) whales, used to capture food by filtering organisms from the water

Basalt   dark volcanic rock, rich in iron and magnesium; characteristic of oceanic basins and mid-ocean ridges

Bay   a partially enclosed ocean inlet

Beach   seaward limit of the shore (between high- and low-tide levels); also the unconsolidated materials that occur there

Benthic   pertaining to the ocean bottom

Benthic boundary layer   a well-mixed layer of water near the ocean bottom (see **Boundary layer**)

Biodiversity   the complexity of life forms on Earth

Biogeochemical cycle   the cycle of movements of elements and other materials through organisms and nonliving systems

Biological pump   the combination of processes that removes nutrients and other substances from surface waters and transports them below the pycnocline into deeper waters, where they eventually return to the surface zone, primarily through upwelling

Bioluminescence   light produced by organisms

Biota   all living things on Earth

Bioturbation   displacement of particles within sediment deposits by biological activity

Bloom   see **Plankton bloom**

Bolus   a body of water with properties distinctly different from those of the surrounding waters

Boundary layer   a thin layer of fluid near an interface (such as the sea surface or the ocean bottom) that exhibits special characteristics, such as high levels of turbulence or Ekman processes; fluid molecules next to the interface do not move with the rest of the fluid

Brackish water   water that is too salty to drink but less salty than average seawater

Breaker   a wave that breaks when it arrives at the shore; the lower part loses speed because of friction with the bottom; the top moves faster and eventually falls over, forming a curl

Brine   a saltwater solution with salt concentration greater than that of normal seawater (35 parts per thousand)

Calcareous sediment   sediment containing more than 30% calcium carbonate, normally as calcareous shells of organisms

Calorie   the amount of heat required to raise the temperature of 1 g of pure water by 1°C

Capillary waves   small waves, with wavelengths less than 1.7 cm (0.67 in.); surface tension is the principal restoring force

Carbohydrates   organic compounds consisting of carbon, hydrogen, and oxygen; sugars and starches are carbohydrates

Carnivore   an animal that feeds on other animals

Cation   an ion with a deficiency of electrons

Chaos   the attribute of complex systems by which the outcome of a set of processes is critically dependent on minute variations in the initial conditions; thus, the outcome is not predictable by standard methods

Chemosynthesis   a process producing energy-rich organic compounds from inorganic constituents, using energy from chemical oxidation rather than energy from the Sun

Clay   very fine-grained sediments; particles less than 4 μm (160 μin.), in diameter

Climate   weather conditions averaged over a long period, typically 30 years

Climate modeling use of computers, complex computer programs, and large sets of climatic data to simulate climatic changes over long periods of time

Coastal current a current paralleling the shoreline; may be caused by tides, winds, or fresh water discharges

Coastal plain a low-lying plain along a coast; usually a landward extension of a continental shelf

Cold event colder-than-average sea-surface temperatures in the central or eastern equatorial Pacific regions (opposite to El Niño conditions; also called La Niña)

Colloids very small particles (less than 0.5 μm, or 20μin., in diameter) that are kept in suspension by random motions of water molecules

Comet an aggregation of rocks and dust with a gaseous cloud

Computer an electronic device for performing calculations or compiling or correlating data

Computer program a set of instructions that tells a computer what to do

Conservative property any property whose value does not change within a particular set of processes; temperature and salinity of seawater are conservative properties below the pycnocline

Continental drift theory a theory that all continents were once connected together in various configurations and subsequently drifted apart

Continental glacier a large ice sheet that covers a large part of a continent

Continental margin a zone along the edge of a continent; includes continental shelf, slope, and rise

Continental rise a region of gently sloping ocean floor between the base of the continental slope and the deep-ocean floor or abyssal plain

Continental shelf a shallow, submerged extension of a coastal plain

Contour a line connecting points of equal value, such as equal temperature or equal elevation

Convection movement of water, air, or Earth materials caused by differences in temperature; Earth's interior is primarily cooled by convection in the mantle

Convergence movement of a substance toward an area

Copepod a small, planktonic herbivore; dominates in many surface-ocean waters

Coral reef a limestone structure, built through growth of corals (animals) and coralline algae (plants), that cements a reef together

Coriolis effect an effect resulting from Earth's rotation, which apparently deflects moving masses to the right in the Northern Hemisphere (and to the left in the Southern Hemisphere) when viewed in the direction of motion, standing on Earth

Critical depth the depth at which photosynthesis equals respiration for a water column

Crust the upper solid shell of Earth; less dense than the underlying mantle

Cyanobacteria very small photosynthetic organisms that are prevalent in waters low in dissolved nutrients

Database a set of data stored so that it can be used by many users without changing its contents

Decomposition breakdown of organic matter into its constituents

Deep-ocean circulation currents, primarily driven by density differences, arising from chilled surface waters and salt from freezing sea ice; also known as thermohaline circulation

Delta land formed by sediment deposits where a river enters an ocean or a large lake

Density mass per unit volume of a substance; also mass per unit area of organisms

Desalination removal of salt from seawater so that it can be used for drinking, for crop irrigation, or for other purposes

Desorption release of chemicals from particle surfaces

Determination direct measurement of a variable

Detritus waste products and dead organisms; eaten by detritivores

Diatom a microscopic autotrophic organism, characterized by being enclosed in two siliceous shells

Diel occurring on a 24-hour cycle

Diurnal tide a tide with a single high and low per day

Divergence   horizontal fluid flows away from an area or line; results in upwelling in water; also occurs in plate movements and in the atmosphere

Doldrums   a band of light and variable winds near the equator

Downwelling   downward movement of a fluid; examples include downwelling of magma in the mantle and downwelling of waters near a coast

Earthquake   sudden breaking and movement of pieces of Earth's crust and the resulting release of energy

Eastern boundary current   a broad, shallow, slow-moving current on the eastern side of an ocean basin

Ebb current   outgoing tidal current

Ecology   the study of living things and the environments in which they live

Ecosystem   an ecological unit including organisms and the environment from which they derive energy and materials needed for maintenance and growth

Ekman Spiral   idealized representation of changing directions and speeds of water flows below the surface caused by a steady wind blowing across an infinitely deep water mass without boundaries

Ekman transport   horizontal average motions of water within a boundary layer, influenced by turbulence or Earth's rotation

El Niño   warming of eastern Pacific equatorial waters and along the Pacific coast of South America; occurs every 4 to 5 years and typically lasts about 12 to 18 months; affects weather worldwide; oceanic component of ENSO (El Niño–Southern Oscillation)

ENSO   full cycle of the Southern Oscillation, including warming of sea-surface temperatures as well as cooling, in comparison with long-term average conditions

Epibenthic   living above sediments, directly on the bottom

Equilibrium   a state of balance between two opposing forces

Estimation   value of one variable as derived from determination of other variables; for example, salinity can be derived from determining the amount of chlorine in a seawater sample

Estuary   a partially enclosed embayment where salt and fresh waters mix

Eukaryote   a complex cell with a true nucleus and other organelles

Euphotic zone   portion of the near-surface ocean having sufficient sunlight to support net photosynthesis; also called the photic zone

Eutrophication   high levels of nutrients caused by increased inputs of nutrients, usually from municipal sewage discharge, agricultural runoff, or windborne pollutant transport

Evaporation   conversion of a liquid to a gas

Evaporite   sedimentary rock produced by evaporation of seawater, which precipitates its salts

Extinction   the dying out of a species; also, great reduction in geographic range of occurrence

Extratropical cyclone   a major storm that does not originate in the tropics; sometimes called a northeaster

Fecal pellets   aggregates of particulate material defecated by animals

Fetch   an area where waves are generated by a wind of constant direction and speed

Fishery   the taking of fish or other marine organisms for human use

Fjord   a deep, drowned valley formed by glacial action

Flocculation (or coagulation)   a process by which small particles coalesce to form larger aggregates as a result of electrical charges on particle surfaces

Flood current   an incoming tidal current

Fluvial   having to do with rivers

Food chain   a simple set of feeding relationships

Food web   a complicated set of feeding relationships among organisms in an ecosystem

Foraminifera   one-celled animals, usually with carbonate shells

Fractal   a self-similar object that seems equally complicated no matter how closely one examines it

Fracture zone   a zone of dislocation of a mid-ocean ridge, usually marked by transverse ridges and troughs

Fringing reef   a coral reef immediately adjacent to a shoreline

Front   a sharp, horizontal boundary marked by sharp changes in air or water properties

Gaia hypothesis   a theory that Earth and all its life forms and interacting systems behave as one large organism

Geosphere   Earth, its atmosphere, and its ocean

Geostationary satellite   a satellite in an orbit around Earth such that it stays above exactly the same place on Earth's surface

Geostrophic current   a current that takes place over large spaces and long time periods, wherein pressure gradients (or gravity) balance the Coriolis effect associated with Earth's rotation

Geothermal energy   heat from inside Earth, sometimes used for human needs

Glacial period   an interval of cold climate that occurs when continental ice sheets expand

Global Positioning System (GPS)   a system of Earth-orbiting satellites that can provide accurate locations and elevations to aircraft, ships, or buoys

Global warming   an increase in the average temperature of Earth's surface resulting from changes in the atmosphere, caused by the release of industrial gases and especially carbon dioxide from burning fossil fuels

Gondwanaland   part of a major supercontinent (Pangea) that broke up about 200 million years ago when the present ocean basins began to form

Granite   igneous rock (formed from magma) that is rich in light elements; typical of continental crust

Graph   a two-dimensional illustration showing a relationship between variables or properties

Great circle   part or all of a circle whose center is the same as Earth's

Greenhouse effect   warming of Earth's atmosphere produced by the presence in the atmosphere of carbon dioxide and other gases, such as methane, which trap heat reflected from Earth's surface; also used to describe increased warming of Earth's atmosphere caused by increased concentrations of radiatively active gases released by human activities

Guyot   a flat-topped seamount, usually formed by a volcano whose top is eroded by weathering and wave action

Gyre   a nearly circular system of surface currents in a major ocean basin

Half-life   time required for half of a given amount of a radioactive substance to decay

Heat budget   balance between energy from the Sun and its radiation back to space

Heat capacity   the amount of heat required to raise the temperature of 1 gram of material by 1°C

Herbivore   an animal that feeds only on plants

High   a weather system in which atmospheric pressure is higher than the surrounding atmosphere

Horse latitudes   ocean regions around 30°N and 30°S where semipermanent high-pressure systems exist

Hot spot (volcanic center)   an anomalously hot region in the mantle below the lithosphere; a source of magma and volcanic activity in the overlying lithospheric plate as it moves across the hot spot

Hurricane   an intense storm in which wind speeds exceed 120 km/hr (74 mi/hr)

Hydrogen bond   a bond formed by the weak attraction between a hydrogen atom in one molecule and a negatively charged atom in another molecule

Hydrogenous sediment   materials precipitated from seawater by chemical processes

Hydrologic cycle   the cycle of water moving among the ocean, rivers, lakes, the atmosphere, and ice sheets; also called the water cycle

Hydrothermal circulation   circulation of hot water in the crust; water is heated by passing through hot, newly formed rocks

Hydrothermal vent   discharge of hot seawater that has flowed through hot, crustal rocks

Ice age   any geological period in which glaciers persisted over large land areas and sea ice over large ocean areas

Iceberg a large piece of a glacier that has broken off and floats in the ocean; icebergs from mountain glaciers (western Greenland) form jagged blocks, whereas those from floating ice shelves (Antarctica) form nearly flat-topped blocks

Ichthyoplankton larval fish too small to move against currents

Industrial Revolution the change from an animal-powered agricultural society to a machine-powered, fossil-fueled industrial society, beginning around 1850 in Europe and North America

Infauna organisms living or burrowing in sediment deposits

Insolation incoming energy from the Sun

Interface a boundary between two substances or waters with marked differences in density or in some other property

Interglacial period a period between ice ages when temperatures are warmer than average

Internal wave a wave occurring within a fluid at a density interface

International date line a line along the 180° meridian of longitude (approximately) where the clock changes; crossing westbound you go ahead one day in time, whereas crossing eastbound you go back one day in time

Interstitial between sediment particles

Intertidal region the seabed lying below high-tide levels but exposed at low tide

Intertropical Convergence Zone (ITCZ) a belt near the equator where prevailing Trade Winds come together, forming a semipermanent low-pressure region

Invertebrates animals lacking backbones and often having external shells or skeletons

Ion a positively or negatively charged atom

Ionic bond a bond formed between two differently charged atoms or molecules

Island arc a chain of volcanic islands lying over a subduction zone, usually near a trench

Isobaths a contour of constant water depth

Isostasy rising or falling of Earth's crust resulting from changes in its density

Isotherm a line joining places of equal temperature

Jet stream a high-speed air stream, about 12 km (7 mi) high; in mid-latitudes, it separates cold air to the north from warmer air to the south

Kelp large brown algae that grow attached to the bottom

Labile readily decomposed; usually refers to organic compounds readily decomposed to inorganic constituents

La Niña part of the El Niño cycle when tropical Pacific waters are unusually cold; also called a cold event

Latent heat heat given off (or taken up) when a substance solidifies (or melts)

Laurasia part of a major supercontinent (Pangea) that broke up when the present ocean basins formed about 200 million years ago; it included North America, Europe, and much of Asia

Lava molten rock erupted by a volcano or fissure

Law of the Sea a comprehensive treaty, negotiated by a UN conference, that established legal status for maritime boundaries (including Exclusive Economic Zones, or EEZs), claims for ocean- bottom resources, and rights of passage for ships passing through international straits

Leeward downwind, or the side away from the wind

Levee an earthen ridge used for flood control; levees can be formed naturally by flooding rivers

Lithogenous sediment sediment consisting primarily of rock or mineral fragments formed by weathering or volcanic eruptions

Little Ice Age a period of colder-than-normal weather that occurred from about 1400 to about 1850 in northern Europe

Longitude distance (measured in degrees) east or west of the Prime Meridian, which passes through Greenwich, England and is opposite the International date line

Low a weather system where atmospheric pressures are lower than those in the surrounding atmosphere

Lunar day the time elapsed between successive instances of the Moon passing directly overhead; approximately 24 hours, 50 minutes

Magma molten rock from deep in Earth's interior; it forms igneous rocks when it solidifies

Magnetic anomaly an area where Earth's magnetic field is either weaker or stronger than normal; caused by magnetic properties in the underlying crust

Manganese nodule a concentric nodule formed by precipitation of iron and manganese compounds from seawater

Mantle the part of Earth's interior lying between the core and the crust; it makes up about two-thirds of Earth's mass and 80% of its volume

Map a flat representation of Earth's surface or some other feature

Mariculture commercial cultivation of marine organisms, both plants and animals

Maritime boundaries national boundaries that extend under the ocean

Meroplankton organisms whose larval stages are planktonic; adults are usually benthic

Mesoscale processes atmospheric processes ranging in scale from 10 to 300 km (6 to 200 mi)

Metamorphism changes in the chemical and mineral compositions of rocks resulting from high temperatures and pressures

Meteorite an extraterrestrial object that passes through Earth's atmosphere and reaches Earth's surface

Microbial pertaining to microbes (microorganisms)

Micron one-millionth of a meter, one-thousandth of a millimeter

Micronutrients substances needed only in minute quantities by organisms to grow and maintain themselves; for instance, vitamins in humans

Microprocessor an integrated circuit that functions in a computer as the central processing unit

Microwave sounder a satellite-mounted instrument used to determine atmospheric water content between the satellite and Earth's surface

Mid-ocean ridges mid-ocean mountain ranges, occurring in all ocean basins; zones where oceanic crust forms

Migration habitual long-distance travel of animals between feeding and breeding/spawning grounds

Mixed tide tide consisting of two unequal high tides and two unequal low tides per day

Mixing blending of waters having different characteristics into a homogeneous mass

Monsoons seasonally variable winds, common in northern Indian Ocean

Nanoplankton plankton between 2 and 20 µm (80 and 800 µin.) in diameter

Nekton animals able to swim against currents

Nepheloid water cloudy or turbid water containing suspended particles

Neuston organisms living at or near the ocean surface

New production that part of primary production supported by nutrients that have not been recently recycled—i.e., not recycled within the preceding 24 hours

Nonconservative property a property whose value changes within a particular set of processes

Nutrients compounds necessary for growth of primary producers; nitrogen and phosphorus compounds are examples

Ocean color scanner a satellite-borne instrument that determines surface water colors: indicates presence of near-surface chlorophyll and other substances that discolor surface waters

Oceanography the study of the ocean, its evolution, and its life forms, and of the ocean basins, their history, and their structure

Oligotrophic water water having low concentrations of nutrients and few organisms

Omnivore an animal that eats both plants and animals

Ooze deep-ocean sediment containing a significant percentage (>30%) of shells of planktonic organisms

Organelles specialized parts of cells that carry out specific functions, much as organs do in more complex organisms

Outgassing   escape of gases from Earth's interior; thought to be the source of Earth's ocean and atmosphere

Ozone layer   a stratospheric layer, 30 to 40 km (20 to 25 mi) thick, containing high concentrations of ozone (a three-atom form of oxygen), which absorbs incoming harmful ultraviolet rays from the Sun

Ozone pollution   ozone formed near Earth's surface by chemical reactions between pollutants (such as automobile exhaust) induced by incoming solar radiation

Pangea   a supercontinent that existed until about 200 million years ago; it included all present continents arranged in two parts, Laurasia (northern) and Gondwanaland (southern)

Pelagic deposits   deposits that accumulate particle by particle

Pelagic environment   upper few hundred meters of the ocean

Photic zone   well-lighted, near-surface waters

Photosynthesis   process by which chlorophyll-bearing organisms use energy from sunlight to convert water and carbon dioxide into energy-rich compounds and release gaseous oxygen

Phytoplankton   single-celled photosynthetic organisms

Picoplankton   planktonic organisms between 0.2 and 2 µm (8 and 80 µin.) in diameter

Pillow lava   volcanic rocks that cooled in forms shaped somewhat like pillows

Plankton   small organisms unable to swim against currents

Plankton bloom   an unusually high concentration of plankton

Plate convergence zone   area where plates are moved together by plate and mantle movements

Plate divergence zone   area where plates are moved apart by crustal or mantle movements

Plate tectonics   the process by which Earth's crust moves as rigid, large plates, while forming and destroying ocean basins and continents

Polar molecule   a molecule, such as water, in which positive and negative charges are separated

Polar-orbiting satellite   a satellite whose orbit passes over both poles; can be in a Sun-synchronous orbit, which passes over each area of Earth's surface at the same time of the solar day

Pollution   degradation of the environment

Polyp   a sedentary organism with a circular body and usually with tentacles surrounding its mouth

Precision   the difference between one result and the average of several results obtained by the same method; reproducibility

Prevailing wind   a wind that blows from the same direction most of the year

Primary production   the amount of photosynthetically formed organic carbon produced per unit time in a unit volume of seawater

Productivity   the amount of organic carbon produced by a specified class of organisms or a particular trophic level (e.g., phytoplankton, zooplankton; primary, secondary, tertiary) per unit of seawater per unit time

Progressive wave   a wave whose the wave form moves

Prokaryote   a simple cell that has no true nucleus or other cellular organelles; examples are bacteria and blue-gree algae

Proxy indicators   features used to deduce past climatic conditions

Pycnocline   a zone marked by increased density with depth separating surface and deep zones

Radioactivity   decay of unstable forms (isotopes) of chemical elements, which change into more stable forms and release radiation and particles

Radiometer   a device used to measure radiation intensity from the sea surface; sea-surface temperature can be determined from such measurements; may be mounted on Earth-orbiting satellites or on aircraft

Random error   an error resulting from basic limitations in the measuring method—for example, how accurately one can read a thermometer

Recruitment   entry of fisheries species into a harvestable stage

Red tide   a massive bloom of dinoflagellates or other red-colored organisms that discolor surface waters and can release toxins that poison fish and other organisms, including humans

Resource a material used for some purpose, such as human needs; may be living, such as fishes, or nonliving, such as sand or gravel

Respiration the process by which energy-rich foods are transformed into energy and carbon dioxide

Rift valley a trough between two fault zones, which is dropped down lower than surroundings

Ring a relatively small water body (10 to 100 km, or 6 to 60 mi, in diameter) surrounded by strong currents; usually forms when cut off from western boundary currents

Rip current a strong surface current flowing seaward through the surf zone

Salinity the total amount of dissolved salt (in grams) in a kilogram of seawater; expressed in parts per thousand

Salt marsh a low-lying area periodically submerged by tides; usually covered by salt-tolerant grasses and other plants

Sargasso Sea the western portion of the North Atlantic subtropical gyre; it gets it name from the floating Sargassum weed that is brought there by converging surface waters

Scatterometer radar used to measure sea-surface roughness; data can be used to determine average wave height or surface wind speed and direction

Sea an area of sharp-peaked, chaotic waves, usually in stormy regions where wind waves are formed

Seismic tomography a technique for analyzing earthquake waves passing through Earth's interior to determine properties and dimensions of various layers

Semidiurnal tide a tide with two highs and two lows per lunar day

Sessile organism an organism fixed in place so that it cannot move

Siliceous sediment sediment formed primarily from siliceous shells of plants or animals

Silled basin a basin in which exchanges with deep waters outside the basin are inhibited by the presence of a ridge (sill) at the entrance

Southern (Antarctic) Ocean ocean surrounding Antarctica, which forms its southern boundary; bounded on the north by the Antarctic Polar Front (also known as the Antarctic Convergence), at about 50°N in the Atlantic and Indian oceans, but at 60°S in the Pacific

Southern Oscillation see-saw exchange of mass (which affects surface atmospheric pressure) in the atmosphere; the atmospheric component of ENSO

Stability a systemic state in which the system will not change radically when perturbed, or a measure of this tendency

Standing wave a wave whose wave form is stationary; also called a seiche

Storm surge a large wavelike feature caused by a large storm, such as a hurricane, through low atmospheric pressure and strong winds

Subduction a process whereby a plate is dragged under an adjoining plate

Submarine canyon a large canyon cut into a continental margin; usually cut by a river at a time of low sea level

Supercontinent an aggregation of large continental blocks swept together by plate tectonic processes; breaks up when new cycle begins and new ocean basins form

Superplume an exceptionally large plume of molten rock; when such a plume reaches Earth's surface, it forms large volcanic provinces on land and oceanic plateaus on the ocean bottom

Sustainable development economic and social development that meets the needs of present generations without compromising the ability of future generations to meet their own needs

Suture a zone formed when two continents collide

Swell regular, smooth-crested waves; that occur outside areas where wind waves are forming

Synthetic aperture radar side-looking radar that maps sea-surface waves and sea ice; requires special computer processing of signals to achieve high resolution of images

Systematic error an error that results from a basic but unrecognized fault in a measuring method

Tectonic process   a large-scale process that moves Earth's crust horizontally (sea-floor spreading) or vertically (mountain building)

Teleconnections   atmospheric interactions among widely separated regions

Temperature of initial freezing   the temperature at which seawater begins to freeze

Temperature of maximum density   the temperature at which pure water reaches its maximum density

Thermocline   the boundary between the warmer surface zone and colder, underlying waters

Thermohaline   having to do with the temperature and/or salinity of the ocean

Tomography   a technique of using computers to obtain a three-dimensional image of the interior of a body; used to obtain internal images of the ocean's depths or Earth's interior

Trade Winds   prevailing winds that blow mainly from the east in the tropics

Transform fault   a fault with horizontal movements separating two plate segments; may occur at a mid-ocean ridge or may form a plate boundary; plates slide past each other along a transform fault

Trench   a long, narrow, deep depression in the ocean floor, associated with subduction

Trophic levels   groups of organisms defined by feeding preferences

Tsunami   a large sea wave formed by an earthquake or an explosive volcanic eruption (Japanese word)

Turbidity current   dense, mud-rich water that flows along the ocean bottom; can erode the bottom as well as transport large amounts of sediment; forms distinctive sediment deposits called turbidites

Turbulence   random movements in a gas or liquid, making the flow unsteady, containing eddies and other random fluctuations

Typhoon   the name for hurricane in the Pacific Ocean

Upwelling   rising of water to the surface from intermediate depths, usually caused by winds moving surface waters away from an area, causing underlying waters to rise

Virus   an organism that is between a simple microorganism and a complex molecule; an important predator on oceanic microscopic organisms

Viscosity   the property of a liquid to resist flow

Volcanic ash   volcanic rock fragments, usually formed during explosive eruptions

Volcanic center (hot spot)   an area of persistent volcanic activity

Volcano   a mountain formed by prolonged eruptions of lava

Warm event   anomalous warming of equatorial surface waters in the central and eastern Pacific

Water mass   a large mass of water that can be identified by characteristic temperatures and salinities in the deep ocean

Wave   a disturbance of a water surface, usually caused by winds

Weathering   the breakdown of rocks at Earth's surface by chemical processes or by physical processes, such as grinding by a glacier moving over the surface

Western boundary current   a strong surface current on the western side of an ocean basin

Windward   on the side facing the wind

Wind wave   a wave formed by winds

Zooplankton   animals too small to swim against currents

Zooxanthellae   dinoflagellates that live in the tissues of other animals, such as corals; usually brownish in color

Index